LONDON MATHEMATICAL SOCIETY LECTURE NOTE SERIES

Managing Editor: Professor M. Reid, Mathematics Institute, University of Warwick, Coventry, CV4 7AL, United Kingdom

The titles below are available from booksellers, or from Cambridge University Press at www.cambridge.org/mathematics

W0042183

London Mathematical Society Lecture Note Series: 355

Non-equilibrium Statistical Mechanics and Turbulence

JOHN CARDY, GREGORY FALKOVICH AND
KRZYSZTOF GAWEDZKI

Edited by

SERGEY NAZARENKO AND OLEG V. ZABORONSKI

CAMBRIDGE
UNIVERSITY PRESS

CAMBRIDGE
UNIVERSITY PRESS

University Printing House, Cambridge CB2 8BS, United Kingdom

Published in the United States of America by Cambridge University Press, New York

Cambridge University Press is part of the University of Cambridge.

It furthers the University's mission by disseminating knowledge in the pursuit of education, learning and research at the highest international levels of excellence.

www.cambridge.org
Information on this title: www.cambridge.org/9780521715140

First published 2008

A catalogue record for this publication is available from the British Library

Library of Congress Cataloguing in Publication data

ISBN 978-0-521-71514-0 Paperback

Contents

Preface

Understanding of turbulence is one of the most challenging problems of modern mathematical and theoretical physics. Turbulence can be described as a chaotic, highly non-equilibrium state of a non-linear physical system. Defined this way, turbulence embraces a broader class of examples than the chaotic Navier-Stokes flow, - a system for which the concept of turbulence was originally introduced and developed. In particular, turbulence appears as far-from-equilibrium states in plasmas, solids, Bose-Einstein condensates and even in nonlinear optics. The characteristic feature of turbulence is the presence of significant energy exchange between many degrees of freedom, which renders most attempts of perturbative treatment of the problem useless. Still mathematicians, physicists and engineers invest a large effort in understanding of turbulence due to the unprecedented importance of turbulence both for theoretical and applied science.

Thanks to fundamental works of Richardson, Taylor, Kolmogorov and Obukhov, we have a phenomenology of turbulent cascades leading to the famous Kolmogorov spectrum. This theory has been successfully applied to a wide range of turbulent systems. What is still lacking, however, is a fundamental theory of turbulence, which would allow both finding a rigorous mathematical foundation for the Kolmogorov theory and understanding of its limitations. The stepping stones on the way to the construction of such a fundamental theory are the solved models of turbulent phenomena. The list of such models which have been understood theoretically, has been growing steadily over the past 60 years. The examples include the theory of wave turbulence, passive scalar advection, the theory of kinematic magnetic dynamo, Burgers turbulence and constant flux states in cluster-cluster aggregation. The success in solving all these problems came from the intensive use of machinery of non-equilibrium statistical physics, such as kinetic theory, quantum field theory-inspired instanton analysis of rare fluctuations and the theory of non-equilibrium critical phenomena.

There is no doubt that non-equilibrium statistical mechanics will play an increasingly important role in further progress of theoretical turbulence. Unfortunately, the range of tools and methods of non-equilibrium statistical mechanics used in modern turbulence research is so wide and they are developing so fast that there is not a single text book which could introduce a graduate student to this area of research. The goal of the present volume of LMS Lecture Notes is to start filling this gap. The book contains three sets of lecture notes written by world-class experts in the non-equilibrium statistical mechanics, which introduce the reader to ideas, methods and applications of non-equilibrium statistical mechanics to turbulence.

The course by professor Gregory Falkovich (Weizmann Institute of Science) gives an introduction to turbulence from the perspective of a statistical physics.

The course by professor Krzysztof Gawedzki (ENS Lyon) gives a thorough introduction into the problem of passive advection using rigorous methods of statistical physics.

The course by professor John Cardy (Oxford University) introduces the reader to field theoretical methods for non-equilibrium critical phenomena.

All courses are equipped with worked-out exercises illustrating the subtle points raised in the lectures and deepen the reader's understanding of the presented theoretical material.

These lectures have been given at the LMS 2006 Summer School at Warwick as part of Warwick Turbulence Symposium. The lectures have been accompanied by example classes led by course assistants - Alexander Fouxon, Adam Gamsa and Peter Horvai. We are indebted to these gentlemen for their help in preparing the lecture notes, designing and typing the class problems and their solutions.

Sergey Nazarenko and Oleg Zaboronski, September 2007.

1

Gregory Falkovich. Introduction to turbulence theory.

The emphasis of this short course is on fundamental properties of developed turbulence, weak and strong. We shall be focused on the degree of universality and symmetries of the turbulent state. We shall see, in particular, which symmetries remain broken even when the symmetry-breaking factor goes to zero, and which symmetries, on the contrary, emerge in the state of developed turbulence.

1.1 Introduction

Turba is Latin for crowd and "turbulence" initially meant the disordered movements of large groups of people. Leonardo da Vinci was probably the first to apply the term to the random motion of fluids. In 20th century, the notion has been generalized to embrace far-from-equilibrium states in solids and plasma. We now define turbulence as a state of a physical system with many interacting degrees of freedom deviated far from equilibrium. This state is irregular both in time and in space and is accompanied by dissipation.

We consider here developed turbulence when the scale of the externally excited motions deviate substantially from the scales of the effectively dissipated ones. When fluid motion is excited on the scale L with the typical velocity V, developed turbulence takes place when the Reynolds number is large: $Re = VL/\nu \gg 1$. Here ν is the kinematic viscosity. At large Re, flow perturbations produced at the scale L have their viscous dissipation small compared to the nonlinear effects. Nonlinearity produces motions of smaller and smaller scales until viscous dissipation stops this at a scale much smaller than L so that there is a wide (so-called inertial) interval of scales where viscosity is negligible and nonlinearity dominates. Another example is the system of waves excited on a fluid surface by wind or moving bodies

1

and in plasma and solids by external electromagnetic fields. The state of such system is called wave turbulence when the wavelength of the waves excited strongly differs from the wavelength of the waves that effectively dissipate. Nonlinear interaction excites waves in the interval of wavelengths (called transparency window or inertial interval) between the injection and dissipation scales.

Simultaneous existence of many modes calls for a statistical description based upon averaging either over regions of space or intervals of time. Here we focus on a single-time statistics of steady turbulence that is on the spatial structure of fluctuations in the inertial range. The basic question is that of universality: to what extent the statistics of such fluctuations is independent of the details of external forcing and internal friction and which features are common to different turbulent systems. This quest for universality is motivated by the hope of being able to distinguish general principles that govern far-from-equilibrium systems, similar in scope to the variational principles that govern thermal equilibrium.

Since we generally cannot solve the nonlinear equations that describe turbulence, we try to infer the general properties of turbulence statistics from symmetries or conservation laws. The conservation laws are broken by pumping and dissipation, but both factors do not act directly in the inertial interval. For example, in the incompressible turbulence, the kinetic energy is pumped by a (large-scale) external forcing and is dissipated by viscosity (at small scales). One may suggest that the kinetic energy is transferred from large to small scales in a cascade-like process i.e. the energy flows throughout the inertial interval of scales. The cascade idea (suggested by Richardson in 1921) explains the basic macroscopic manifestation of turbulence: the rate of dissipation of the dynamical integral of motion has a finite limit when the dissipation coefficient tends to zero. For example, the mean rate of the viscous energy dissipation does not depend on viscosity at large Reynolds numbers. Intuitively, one can imagine turbulence cascade as a pipe in wavenumber space that carries energy. As viscosity gets smaller the pipe gets longer but the flux it carries does not change. Formally, that means that the symmetry of the inviscid equation (here, time-reversal invariance) is broken by the presence of the viscous term, even though the latter might have been expected to become negligible in the limit $Re \to \infty$.

One can use the cascade idea to guess the scaling properties of turbulence. For incompressible fluid, the energy flux (per unit mass) ϵ through the given scale r can be estimated via the velocity difference δv measured at that scale as the energy $(\delta v)^2$ divided by the time $r/\delta v$. That gives $(\delta v)^3 \sim \epsilon r$. Of course, δv is a fluctuating quantity and we ought to make statements on its moments or probability distribution $\mathcal{P}(\delta v, r)$. Energy flux constancy fixes the third moment, $\langle (\delta v)^3 \rangle \sim \epsilon r$. It is a natural wish to have turbulence scale invariant in the inertial interval so that $\mathcal{P}(\delta v, r) = (\delta v)^{-1} f[\delta v / (\epsilon r)^{1/3}]$ is expressed via the dimensionless function f of a single variable. Initially, Kolmogorov made even stronger wish for the function f to be universal (i.e. pumping independent). Nature is under no obligation to grant wishes of even great scientists, particularly when it is in a state of turbulence. After hearing Kolmogorov talk, Landau remarked that the moments different from third are nonlinear functions of the input rate and must be sensitive to the precise statistics of the pumping. As we show below, the cascade idea can indeed be turned into an exact relation for the simultaneous correlation function which expresses the flux (third or fourth-order moment depending on the degree of nonlinearity). The relation requires the mean flux of the respective integral of motion to be constant across the inertial interval of scales. We shall see that flux constancy determines the system completely only for a weakly nonlinear system (where the statistics is close to Gaussian i.e. not only scale invariant but also perfectly universal). To describe an entire turbulence statistics of strongly interacting systems, one has to solve problems on a case-by-case basis with most cases still unsolved. Particularly difficult (and interesting) are the cases when not only universality but also scale invariance is broken so that knowledge of the flux does not allow one to predict even the order of magnitude of high moments. We describe the new concept of statistical integrals of motion which allows for the description of system with broken scale invariance. We also describe situations when not only scale invariance is restored but a wider conformal invariance takes place in the inertial interval.

1.2 Weak wave turbulence

It is easiest to start from a weakly nonlinear system. Such is a system of small-amplitude waves. Examples include waves on the water surface, waves in plasma with and without magnetic field, spin waves in magnetics etc. We assume spatial homogeneity and denote a_k the amplitude of the wave with the wavevector $\mathbf{k}$. Considering for a moment wave system as closed (that is without external pumping and dissipation) one can describe it as a Hamiltonian system using wave amplitudes as normal canonical variables — see, for instance, the monograph Zakharov et al 1992. At small amplitudes, the Hamiltonian can be written as an expansion over a_k, where the second-order term describes non-interacting waves and high-order terms determine the interaction†:

$$H = \int \omega_k |a_k|^2 \, d\mathbf{k} \tag{1.1}$$

$$+ \int \left(V_{123} a_1 a_2^* a_3^* + c.c. \right) \delta(\mathbf{k}_1 - \mathbf{k}_2 - \mathbf{k}_3) \, d\mathbf{k}_1 d\mathbf{k}_2 d\mathbf{k}_3 + \mathrm{O}(a^4).$$

The dispersion law ω_k describes wave propagation, $V_{123} = V(\mathbf{k}_1, \mathbf{k}_2, \mathbf{k}_3)$ is the interaction vertex and c.c. means complex conjugation. In the Hamiltonian expansion, we presume every subsequent term smaller than the previous one, in particular, $\xi_k = |V_{kkk} a_k| k^d / \omega_k \ll 1$ — wave turbulence that satisfies that condition is called weak turbulence. Here d is the space dimensionality.

The dynamic equation which accounts for pumping, damping, wave propagation and interaction has the following form:

$$\partial a_k / \partial t = -i \delta H / \delta a_k^* + f_k(t) - \gamma_k a_k. \tag{1.2}$$

Here γ_k is the decrement of linear damping and f_k describes pumping. For a linear system, a_k is different from zero only in the regions of $\mathbf{k}$-space where f_k is nonzero. Nonlinear interaction provides for wave excitation outside pumping regions.

It is likely that the statistics of the weak turbulence at $k \gg k_f$ is close to Gaussian for wide classes of pumping statistics. When the forcing $f_k(t)$ is Gaussian then the statistics of $a_k(t)$ is close to Gaussian as long as nonlinearity is weak. However, in most cases in nature and in the lab, the force is not Gaussian even though its amplitude can be small. It is an open problem to find out under what conditions the wave field is close to Gaussian with $\langle a_k(0) a_{k'}^*(t) \rangle = n_k \exp(-i\omega_k t) \delta(\mathbf{k} + \mathbf{k}')$. This problem actually breaks into two parts. The first one is to solve the linear equation for the waves in the spectral interval of pumping and formulate the criteria on the forcing that guarantee that the cumulants are small for $a_k(t) = \exp(-i\omega_k t - \gamma_k t) \int_0^t f_k(t') \exp(i\omega_k t + \gamma_k t) \, dt'$. The second part is more interesting: even when the pumping-related waves are non-Gaussian, it may well be that as we go in k-space away from pumping (into the inertial interval) the field $a_k(t)$ is getting more Gaussian. Unless we indeed show that, most of the applications of the weak turbulence theory described in this section are in doubt. See also Choi et al 2005.

We consider here and below a pumping by a Gaussian random force statistically isotropic and homogeneous in space, and white in time (see also

---

† For example, for sound one expands the (kinetic plus internal) energy density $\rho v^2 / 2 + E(\rho)$ assuming $v \ll c$ and using $\mathbf{v}_k = \mathbf{k}(a_k - a_{-k}^*)(ck/2\rho_0)^{1/2}$, $\rho_k = k(a_k + a_{-k}^*)(\rho_0/2ck)^{1/2}$.

Sect. 3.1 of John Cardy's course):

$$\langle f_k(t) f^*_{k'}(t') \rangle = F(k)\delta(\mathbf{k} + \mathbf{k}')\delta(t - t') . \tag{1.3}$$

Angular brackets mean spatial average. We assume $\gamma_k \ll \omega_k$ (for waves to be well defined) and that $F(k)$ is nonzero only around some k_f.

As long as we assume the statistics of the wave system to be close to Gaussian, it is completely determined by the pair correlation function. Here we are interesting in the spatial structure which is described by the single-time pair correlation function $\langle a_k(t) a^*_{k'}(t) \rangle = n_k(t)\delta(\mathbf{k} + \mathbf{k}')$. Since the dynamic equation (1.2) contains a quadratic nonlinearity then the time derivative of the second moment, $\partial n_k/\partial t$, is expressed via the third one, the time derivative of the third moment ix expressed via the fourth one etc; that is the statistical description in terms of moments encounters the closure problem. Fortunately, weak turbulence in the inertial interval is expected to have the statistics close to Gaussian so one can express the fourth moment as the product of two second ones. As a result one gets a closed equation (see e.g. Zakharov et al 1992):

$$\frac{\partial n_k}{\partial t} = F_k - \gamma_k n_k + I^{(3)}_k , \qquad I^{(3)}_k = \int (U_{k12} - U_{1k2} - U_{2k1}) \, dk_1 dk_2 \tag{1.4}$$

$$U_{123} = \pi \left[n_2 n_3 - n_1 (n_2 + n_3) \right] |V_{123}|^2 \delta(\mathbf{k}_1 - \mathbf{k}_2 - \mathbf{k}_3)\delta(\omega_1 - \omega_2 - \omega_3) .$$

It is called kinetic equation for waves. The collision integral $I^{(3)}_k$ results from the cubic terms in the Hamiltonian i.e. from the quadratic terms in the equations for amplitudes. It can be *interpreted* as describing three-wave interactions: the first term in the integral (1.4) corresponds to a decay of a given wave while the second and third ones to a confluence with other wave.

The inverse time of nonlinear interaction at a given k can be estimated from (1.4) as $|V(k, k, k)|^2 n(k) k^d / \omega(k)$. We define the dissipation wavenumber k_d as such where this inverse time is comparable to $\gamma(k_d)$ and assume nonlinearity to dominate over dissipation at $k \ll k_d$. As has been noted, wave turbulence appears when there is a wide (inertial) interval of scales where both pumping and damping are negligible, which requires $k_d \gg k_f$, the condition analogous to $Re \gg 1$. This is schematically shown in Fig. 1.

The presence of frequency delta-function in $I^{(3)}_k$ means that in the first order of perturbation theory in wave interaction we account only for resonant processes which conserve the quadratic part of the energy $E = \int \omega_k n_k \, dk = \int E_k dk$. For the cascade picture to be valid, the collision integral has to converge in the inertial interval which means that energy exchange is small between motions of vastly different scales, the property called interaction locality in k-space (see the exercise 1.1 below). Consider now a statistical

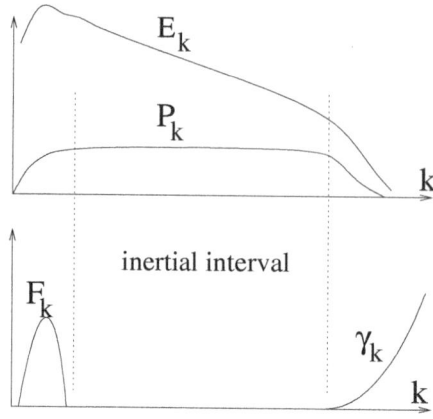

Fig. 1.1. A schematic picture of the cascade.

steady state established under the action of pumping and dissipation. Let us multiply (1.4) by ω_k and integrate it over either interior or exterior of the ball with radius k. Taking $k_f \ll k \ll k_d$, one sees that the energy flux through any spherical surface (Ω is a solid angle), is constant in the inertial interval and is equal to the energy production/dissipation rate ϵ:

$$P_k = \int_0^k k^{d-1} dk \int d\Omega\, \omega_k I_k^{(3)} = \int \omega_k F_k\, d\mathbf{k} = \int \gamma_k E_k\, dk = \epsilon \ . \qquad (1.5)$$

That (integral) equation determines n_k. Let us assume now that the medium (characterized by the Hamiltonian coefficients) can be considered isotropic at the scales in the inertial interval. In addition, for scales much larger or much smaller than a typical scale (like Debye radius in plasma or the depth of the water) the medium is usually scale invariant: $\omega(k) = ck^\alpha$ and $|V(\mathbf{k}, \mathbf{k}_1, \mathbf{k}_2)|^2 = V_0^2 k^{2m} \chi(\mathbf{k}_1/k, \mathbf{k}_2/k)$ with $\chi \simeq 1$. Remind that we presumed statistically isotropic force. In this case, the pair correlation function that describes a steady cascade is also isotropic and scale invariant:

$$n_k \simeq \epsilon^{1/2} V_0^{-1} k^{-m-d} \ . \qquad (1.6)$$

One can show that (1.6), called Zakharov spectrum, turns $I_k^{(3)}$ into zero (see the exercise 1.1 below and Zakharov *et al* 1992).

If the dispersion relation $\omega(k)$ does not allow for the resonance condition $\omega(k_1) + \omega(k_2) = \omega(|\mathbf{k}_1 + \mathbf{k}_2|)$ then the three-wave collision integral is zero and one has to account for four-wave scattering which is always resonant, that is whatever $\omega(k)$ one can always find four wavevectors that satisfy $\omega(k_1) + \omega(k_2) = \omega(k_3) + \omega(k_4)$ and $\mathbf{k}_1 + \mathbf{k}_2 = \mathbf{k}_3 + \mathbf{k}_4$. The collision integral that describes scattering,

$$
I_k^{(4)} = \frac{\pi}{2} \int |T_{k123}|^2 \left[n_2 n_3 (n_1 + n_k) - n_1 n_k (n_2 + n_3) \right] \delta(\mathbf{k} + \mathbf{k}_1 - \mathbf{k}_2 - \mathbf{k}_3)
$$
$$
\times \delta(\omega_k + \omega_1 - \omega_2 - \omega_2) \, d\mathbf{k}_1 d\mathbf{k}_2 d\mathbf{k}_3 \,, \tag{1.7}
$$

conserves the energy and the wave action $N = \int n_k \, d\mathbf{k}$ (the number of waves). Pumping generally provides for an input of both E and N. If there are two inertial intervals (at $k \gg k_f$ and $k \ll k_f$), then there should be two cascades. Indeed, if $\omega(k)$ grows with k then absorbing finite amount of E at $k_d \to \infty$ corresponds to an absorption of an infinitely small N. It is thus clear that the flux of N has to go in opposite direction that is to large scales. A so-called inverse cascade with the constant flux of N can thus be realized at $k \ll k_f$. A sink at small k can be provided by wall friction in the container or by long waves leaving the turbulent region in open spaces (like in sea storms). Two-cascade picture can be illustrated by a simple example with a wave source at $\omega = \omega_2$ generating N_2 waves per unit time and two sinks at $\omega = \omega_1$ and $\omega = \omega_3$ absorbing respectively N_1 and N_3. In a steady state, $N_2 = N_1 + N_3$ and $\omega_2 N_2 = \omega_1 N_1 + \omega_3 N_3$, which gives

$$
N_1 = N_2 \frac{\omega_3 - \omega_2}{\omega_3 - \omega_1}, \qquad N_3 = N_2 \frac{\omega_2 - \omega_1}{\omega_3 - \omega_1} \,.
$$

At a sufficiently large left inertial interval (when $\omega_1 \ll \omega_2 < \omega_3$), the whole energy is absorbed by the right sink: $\omega_2 N_2 \approx \omega_3 N_3$. Similarly, at $\omega_3 \gg \omega_2 > \omega_1$, we have $N_1 \approx N_2$, i.e. the wave action is absorbed at small ω.

The collision integral $I_k^{(3)}$ involved products of two n_k so that flux constancy required $E_k \propto \epsilon^{1/2}$ while for the four-wave case $I_k^{(4)} \propto n^3$ gives $E_k \propto \epsilon^{1/3}$. In many cases (when there is a complete self-similarity) that knowledge is sufficient to obtain the scaling of E_k from a dimensional reasoning without actually calculating V and T. For example, short waves on a deep water are characterized by the surface tension σ and density ρ so the dispersion relation must be $\omega_k \sim \sqrt{\sigma k^3 / \rho}$ which allows for the three-wave resonance and thus $E_k \sim \epsilon^{1/2} (\rho \sigma)^{1/4} k^{-7/4}$. For long waves on a deep water, the surface-restoring force is dominated by gravity so that the gravity acceleration g replaces σ as a defining parameter and $\omega_k \sim \sqrt{gk}$. Such dispersion law does not allow for the three-wave resonance so that the dominant interaction is four-wave scattering which permits two cascades. The direct energy cascade corresponds to $E_k \sim \epsilon^{1/3} \rho^{2/3} g^{1/2} k^{-5/2}$. The inverse cascade carries the flux of N which we denote Q, it has the dimensionality $[Q] = [\epsilon]/[\omega_k]$ and corresponds to $E_k \sim Q^{1/3}\rho^{2/3}g^{2/3}k^{-7/3}$.

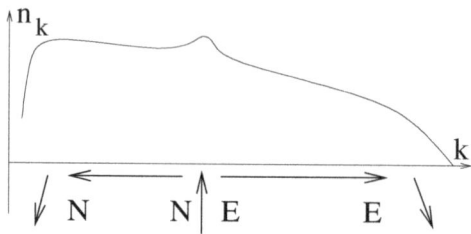

Fig. 1.2. Two cascades under four-wave interaction.

Under a weakly anisotropic pumping, stationary spectrum acquires a small stationary weakly anisotropic correction $\delta n(\mathbf{k})$ such that $\delta n(\mathbf{k})/\mathbf{n_0}(\mathbf{k}) \propto \omega(\mathbf{k})/k$ (see exercise 2.2). The degree of anisotropy increases with k for waves with the decay dispersion law. That is the spectrum of the weak turbulence generated by weakly anisotropic pumping is getting more anisotropic as we go into the inertial interval of scales. We see that the conservation of the second integral (momentum) can lead to the non-restoration of symmetry (isotropy) in the inertial interval.

Since the statistics of weak turbulence is near Gaussian, it is completely determined by the pair correlation function, which is in turn determined by the respective flux (or fluxes). We thus conclude that weak turbulence is perfectly universal: deep in the inertial interval it "forgets" all the properties of pumping except the flux value.

1.3 Strong wave turbulence

Weak turbulence theory breaks down when the wave amplitudes are large enough (so that $\xi_k \geq 1$). We need special consideration also in the particular case of the linear (acoustic) dispersion relation $\omega(k) = ck$ for arbitrarily small amplitudes (as long as the Reynolds number remains large). Indeed, there is no dispersion of wave velocity for acoustic waves so that waves moving at the same direction interact strongly and produce shock waves when viscosity is small. Formally, there is a singularity due to coinciding arguments of delta-functions in (1.4) (and in the higher terms of perturbation expansion for $\partial n_k/\partial t$), which is thus invalid at however small amplitudes.

Still, some features of the statistics of acoustic turbulence can be understood even without a closed description. We discuss that in a one-dimensional case which pertains, for instance, to sound propagating in long pipes. Since weak shocks are stable with respect to transversal perturbations (Landau and Lifshits 1987), quasi one-dimensional perturbations may propagate in 2d and 3d as well. In the reference moving with the sound velocity, the weakly compressible 1d flows ($u \ll c$) are described by the Burgers equation (Landau and Lifshits 1987, E et al 1997, Frisch and Bec 2001):

$$u_t + uu_x - \nu u_{xx} = 0 . \qquad (1.8)$$

Burgers equation has a propagating shock-wave solution $u = 2v\{1 + \exp[v(x - vt)/\nu]\}^{-1}$ with the energy dissipation rate $\nu \int u_x^2\, dx$ independent of ν. The shock width ν/v is a dissipative scale and we consider acoustic turbulence produced by a pumping correlated on much larger scales (for example, pumping a pipe from one end by frequencies much less than cv/ν). After some time, it will develop shocks at random positions. Here we consider the single-time statistics of the Galilean invariant velocity difference $\delta u(x,t) = u(x,t) - u(0,t)$. The moments of δu are called structure functions $S_n(x,t) = \langle[u(x,t) - u(0,t)]^n\rangle$. Quadratic nonlinearity relates the time derivative of the second moment to the third one:

$$\frac{\partial S_2}{\partial t} = -\frac{\partial S_3}{3\partial x} - 4\epsilon + \nu\frac{\partial^2 S_2}{\partial x^2} . \qquad (1.9)$$

Here $\epsilon = \nu\langle u_x^2 \rangle$ is the mean energy dissipation rate. Equation (1.9) describes both a free decay (then ϵ depends on t) and the case of a permanently acting pumping which generates turbulence statistically steady at scales less than the pumping length. In the first case, $\partial S_2/\partial t \simeq S_2 u/L \ll \epsilon \simeq u^3/L$ (where L is a typical distance between shocks) while in the second case $\partial S_2/\partial t = 0$ so that $S_3 = 12\epsilon x + \nu\partial S_2/\partial x$.

Consider now limit $\nu \to 0$ at fixed x (and t for decaying turbulence). Shock dissipation provides for a finite limit of ϵ at $\nu \to 0$ then

$$S_3 = -12\epsilon x . \qquad (1.10)$$

This formula is a direct analog of (1.5). Indeed, the Fourier transform of (1.9) describes the energy density $E_k = \langle|u_k|^2\rangle/2$ which satisfies the equation $(\partial_t - \nu k^2)E_k = -\partial P_k/\partial k$ where the k-space flux

$$P_k = \int_0^k dk' \int_{-\infty}^\infty dx\, S_3(x)k' \sin(k'x)/24 .$$

It is thus the flux constancy that fixes $S_3(x)$ which is universal that is determined solely by ϵ and depends neither on the initial statistics for decay nor on the pumping for steady turbulence. On the contrary, other structure functions $S_n(x)$ are not given by $(\epsilon x)^{n/3}$. Indeed, the scaling of the structure functions can be readily understood for any dilute set of shocks (that is when shocks do not cluster in space) which seems to be the case both for smooth initial conditions and large-scale pumping in Burgers turbulence. In this case, $S_n(x) \sim C_n|x|^n + C_n'|x|$ where the first term comes from the regular (smooth) parts of the velocity (the right x-interval in Fig. 1.3) while the second comes from $O(x)$ probability to have a shock in the interval x. The scaling exponents, $\xi_n = d\ln S_n / d\ln x$, thus behave as follows: $\xi_n = n$ for $n \leq 1$ and $\xi_n = 1$ for $n > 1$. That means that the probability density

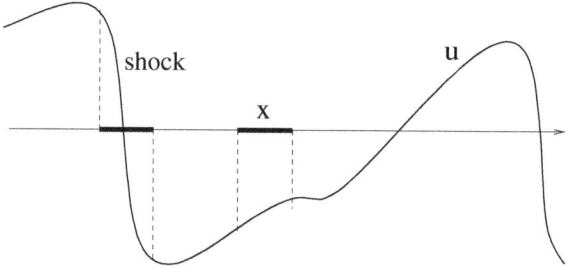

Fig. 1.3. Typical velocity profile in Burgers turbulence.

function (PDF) of the velocity difference in the inertial interval $P(\delta u, x)$ is not scale-invariant, that is the function of the re-scaled velocity difference $\delta u / x^a$ cannot be made scale-independent for any a. Simple bi-modal nature of Burgers turbulence (shocks and smooth parts) means that the PDF is actually determined by two (non-universal) functions, each depending of a single argument: $P(\delta u, x) = \delta u^{-1} f_1(\delta u/x) + x f_2(\delta u/u_{rms})$. Breakdown of scale invariance means that the low-order moments decrease faster than the high-order ones as one goes to smaller scales, i.e. the smaller the scale the more probable are large fluctuations. In other words, the level of fluctuations increases with the resolution. When the scaling exponents ξ_n do not lie on a straight line, this is called an anomalous scaling since it is related again to the symmetry (scale invariance) of the PDF broken by pumping and not restored even when $x/L \to 0$.

As an alternative to the description in terms of structures (shocks), one can relate the anomalous scaling in Burgers turbulence to the additional integrals of motion. Indeed, the integrals $E_n = \int u^{2n}\,dx/2$ are all conserved by the inviscid Burgers equation. Any shock dissipates the finite amount of

E_n at the limit $\nu \to 0$ so that similarly to (1.10) one denotes $\langle \dot{E}_n \rangle = \epsilon_n$ and obtains $S_{2n+1} = -4(2n+1)\epsilon_n x/(2n-1)$ for integer n. We thus conclude that the statistics of velocity differences in the inertial interval depends on the infinitely many pumping-related parameters, the fluxes of all dynamical integrals of motion.

Note that $S_2(x) \propto |x|$ corresponds to $E(k) \propto k^{-2}$, since every shock gives $u_k \propto 1/k$ at $k \ll v/\nu$, that is the energy spectrum is determined by the type of structures (shocks) rather than by energy flux constancy. That is Burgers turbulence demonstrates the universality of a different kind: the type of structures that dominate turbulence (here, shocks) is universal while the statistics of their amplitudes depends on pumping. Similar ideas were suggested for other types of strong wave turbulence assuming them to be dominated by different structures. Weak wave turbulence, being a set of weakly interacting plane waves, can be studied uniformly for different systems (Zakharov et al 1992). On the contrary, when nonlinearity is strong, different structures appear. Broadly, one distinguishes conservative structures (like solitons and vortices) from dissipative structures which usually appear as a result of finite-time singularity of the non-dissipative equations (like shocks, light self-focussing or wave collapse).

For example, an envelope of a spectrally narrow wave packets is described by the Nonlinear Schrödinger Equation ,

$$i\Psi_t + \Delta\Psi + T|\Psi|^2\Psi = 0 . \tag{1.11}$$

This equation also describes Bose-Einstein condensation (then it is usually called Gross-Pitaevsky equation). Weak turbulence is determined by $|T|^2$ and is the same both for $T < 0$ (wave repulsion) and $T > 0$ (wave attraction). Inverse cascade tends to produce a uniform condensate $\Psi(k = 0) = A$. At high levels of nonlinearity, different signs of T correspond to dramatically different physics. At $T < 0$ the condensate is stable, it renormalizes the linear dispersion relation from $\omega_k = k^2$ to the Bogolyubov form $\omega_k^2 = k^4 - 2TA^2k^2$. That dispersion relation is close to acoustic at small k, it allows for three-wave interactions. The resulting over-condensate turbulence is a mixture of phonons, solitons, kinks and vortices, I shall comment briefly on its properties at the end of the course. On the contrary, the condensate and sufficiently long waves are unstable at $T > 0$; that instability leads to wave collapse at $d = 2, 3$ with the energy being fast transferred from large to small scales where it dissipates (Dyachenko et al 1992). No analytic theory is yet available for such strong turbulence.

Nonlinearity parameter $\xi(k)$ generally depends on k so that there may exist weakly turbulent cascade until some k_* where $\xi(k_*) \sim 1$ and strong turbulence beyond this wavenumber, that is weak and strong turbulence can coexist in the same system. Presuming that some mechanism (for instance, wave breaking) prevents appearance of wave amplitudes that correspond to $\xi_k \gg 1$, one may suggest that some cases of strong turbulence correspond to the balance between dispersion and nonlinearity local in k-space so that $\xi(k)$ =const throughout its domain in k-space. That would correspond to the spectrum $E_k \sim \omega_k^3 k^{-d}/|V_{kkk}|^2$ which is ultimately universal that is independent even of the flux (only the boundary k_* depends on the flux). For gravity waves, this gives $E_k = \rho g k^{-3}$, the same spectrum one obtains presuming wave profile to have cusps (another type of dissipative structure leading to whitecaps in stormy sea (Phillips 1977, Kuznetsov 2004). It is unclear if such flux-independent spectra are realized.

1.4 Incompressible turbulence

Incompressible fluid flow is described by the Navier-Stokes equation

$$\partial_t \mathbf{v}(\mathbf{r},t) + \mathbf{v}(\mathbf{r},t) \cdot \nabla \mathbf{v}(\mathbf{r},t) - \nu \nabla^2 \mathbf{v}(\mathbf{r},t) = -\nabla p(\mathbf{r},t), \quad \text{div } \mathbf{v} = 0. \quad (1.12)$$

See Lecture 1 of the Gawędzki course for more details on this equation. We are again interested in the structure functions $S_n(\mathbf{r},t) = \langle[(\mathbf{v}(\mathbf{r},t)-\mathbf{v}(0,t)) \cdot \mathbf{r}/r]^n\rangle$ and consider distance r smaller than the force correlation scale for a steady case and smaller than the size of turbulent region for a decay case.

1.4.1 Three dimensional turbulence

. We treat first the three-dimensional case. Similar to (1.9), one can derive the Karman-Howarth relation between S_2 and S_3 (see Landau and Lifshits 1987):

$$\frac{\partial S_2}{\partial t} = -\frac{1}{3r^4}\frac{\partial}{\partial r}(r^4 S_3) + \frac{4\epsilon}{3} + \frac{2\nu}{r^4}\frac{\partial}{\partial r}\left(r^4 \frac{\partial S_2}{\partial r}\right). \quad (1.13)$$

Here $\epsilon = \nu \langle(\nabla \mathbf{v})^2\rangle$ is the mean energy dissipation rate. Neglecting time derivative (which is zero in a steady state and small comparing to ϵ for decaying turbulence) one can multiply (1.13) by r^4 and integrate: $S_3(r) = -4\epsilon r/5 + 6\nu dS_2(r)/dr$. Kolmogorov considered the limit $\nu \to 0$ for fixed r and *assumed* nonzero limit for ϵ which gives the so-called 4/5 law (Kolmogorov 1941, Landau and Lifshits 1987, Frisch 1995):

$$S_3 = -\frac{4}{5}\epsilon r. \quad (1.14)$$

Similar to (1.5,1.10), this relation means that the kinetic energy has a constant flux in the inertial interval of scales (the viscous scale η is defined by $\nu S_2(\eta) \simeq \epsilon \eta^2$). Let us stress that this flux relation is built upon the assumption that the energy dissipation rate ϵ has a nonzero limit at vanishing viscosity. Since the input rate can be independent of viscosity, this is the assumption needed for an existence of a steady state at the limit: no matter how small the viscosity, or how high the Reynolds number, or how extensive the scale-range participating in the energy cascade, the energy flux is expected to remain equal to that injected at the stirring scale. Unlike compressible (Burgers) turbulence, here we do not know the form of the specific singular structures that are supposed to provide non-vanishing dissipation in the inviscid limit (as shocks waves do). Experimental data show, however, that the dissipation rate is indeed independent of the Reynolds number when $Re \gg 1$. Historically, persistence of the viscous dissipation in the inviscid limit (both in compressible and incompressible turbulence) is the first example of what is now called "anomaly" in theoretical physics: a symmetry of the equation (here, time-reversal invariance) remains broken even as the symmetry-breaking factor (viscosity) becomes vanishingly small (see e.g. Falkovich and Sreenivasan 2006). If one screens a movie of steady turbulence backwards, we can tell that something is indeed wrong!

The law (1.14) shows that the third-order moment is universal, i.e. it does not depend on the details of the turbulence production but is determined solely by the mean energy dissipation rate. The rest of the structure functions have never been derived. Kolmogorov (1941) and also Heisenberg, von Weizsacker and Onsager *presumed* the pair correlation function to be determined only by ϵ and r which would give $S_2(r) \sim (\epsilon r)^{2/3}$ and the energy spectrum $E_k \sim \epsilon^{2/3} k^{-5/3}$. Experiments suggest that $\zeta_n = d\ln S_n / d\ln r$ lie on a smooth concave curve sketched in Fig. 1.4. While ζ_2 is close to 2/3 it has to be a bit larger because experiments show that the slope at zero $d\zeta_n/dn$ is larger than 1/3 while $\zeta(3) = 1$ in agreement with (1.14). Like in Burgers, the PDF of velocity differences in the inertial interval is not scale invariant in the 3d incompressible turbulence. So far, nobody was able to find an explicit relation between the anomalous scaling for 3d Navier-Stokes turbulence and either structures or additional integrals of motion.

While not exact, the Kolomogorov's approximation $S_2(\eta) \simeq (\epsilon \eta)^{2/3}$ can be used to estimate the viscous scale: $\eta \simeq L Re^{-3/4}$. The number of degrees of freedom involved into 3d incompressible turbulence can thus be roughly estimated as $N \sim (L/\eta)^3 \sim Re^{9/4}$. That means, in particular, that detailed numerical simulation of water or oil pipe flows ($Re \sim 10^4 \div 10^7$) or turbulent cloud ($Re \sim 10^6 \div 10^9$) is beyond the reach of today (and possibly

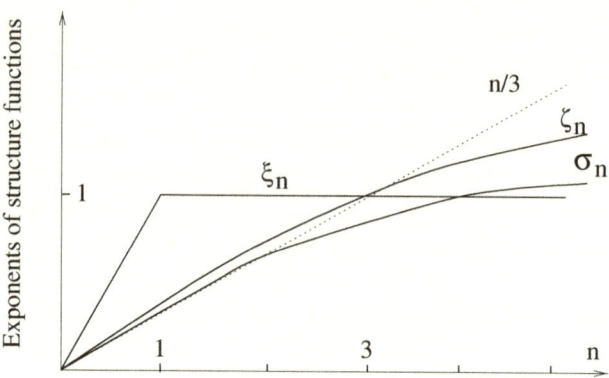

Fig. 1.4. The scaling exponents of the structure functions ξ_n for Burgers, ζ_n for 3d Navier-Stokes and σ_n for the passive scalar. The dotted straight line is $n/3$.

tomorrow) computers. To calculate correctly at least the large-scale part of the flow, it is desirable to have some theoretical model to parameterize the small-scale motions. Here, the main obstacle is our lack of qualitative understanding and quantitative description of how turbulence statistics changes with the scale. This breakdown of scale invariance in the inertial range is another example of anomaly (effect of pumping scale does not disappear even at the limit $r/L \to 0$). Such an anomalous (or multi-fractal) scaling, is an important feature of turbulence, and sets it apart from the usual critical phenomena: one needs to work out the behavior of moments of each order independently rather than get it from dimensional analysis. Anomalous scaling in turbulence is such that $\zeta_{2n} < n\zeta_2$ so that S_{2n}/S_2^n for $n > 2$ increases as $r \to 0$. The relative growth of high moments means that strong fluctuations become more probable as the scales become smaller. Its practical importance is that it limits our ability to produce realistic models for small-scale turbulence.

Since we know neither the structures nor the extra conservation laws that are responsible for an anomalous scaling in the 3d incompressible turbulence, then, to get some qualitative understanding of this very complicated problem, we now pass to another (no less complicated) problem of 2d turbulence. That latter problem will motivate us to consider passive scalar turbulence, which will, in particular, teach us a new concept of statistical conservation laws that will shed some light on 3d turbulence too.

1.4.2 Two-dimensional Turbulence

Large-scale motions in shallow fluid can be approximately considered two-dimensional. When the velocities of such motions are much smaller than the velocities of the surface waves and the velocity of sound, such flows can be considered incompressible. Their description is important for understanding atmospheric and oceanic turbulence at the scales larger than the atmosphere height and the ocean depth. Vorticity $\omega = curl\,\mathbf{v}$ is a scalar in a two-dimensional flow. It is advected by the velocity field and dissipated by viscosity. Taking $curl$ of the Navier-Stokes equation one gets

$$d\omega/dt = \partial_t\omega + (\mathbf{v}\cdot\nabla)\omega = \nu\nabla^2\omega \ . \tag{1.15}$$

Two-dimensional incompressible inviscid flow just transports vorticity from place to place and thus conserves spatial averages of any function of vorticity, $\Omega_n \equiv \int \omega^n d\mathbf{r}$. In particular, we now have the second quadratic inviscid invariant (in addition to energy) which is called enstrophy: $\Omega_2 = \int \omega^2\,d\mathbf{r}$. Since the spectral density of the energy is $|\mathbf{v}_k|^2/2$ while that of the enstrophy is $|\mathbf{k}\times\mathbf{v}_k|^2$ then (similarly to the cascades of E and N in wave turbulence under four-wave interaction) one expects that the direct cascade (towards large k) is that of enstrophy while the inverse cascade is that of energy, as was suggested by Kraichnan (1967). What about other Ω_n? The intuition developed so far might suggest that the infinity of dynamical conservation laws must bring about anomalous scaling. As we shall see, turbulence never fails to defy intuition.

1.4.3 Passive Scalar Turbulence

Before discussing vorticity statistics in two-dimensional turbulence, we describe a similar yet somewhat simpler problem of passive scalar turbulence which allows one to introduce the necessary notions of Lagrangian description of the fluid flow. Consider a scalar quantity $\theta(\mathbf{r}, t)$ which is subject to molecular diffusion and advection by the fluid flow but has no back influence on the velocity (i.e. passive):

$$d\theta/dt = \partial_t\theta + (\mathbf{v}\cdot\nabla)\theta = \kappa\nabla^2\theta \ . \tag{1.16}$$

Here κ is molecular diffusivity. The examples of passive scalar are smoke in the air, salinity in the water and temperature when one can neglect thermal convection. Without viscosity and diffusion, ω and θ behave in the same way in the same 2d flow — they are both Lagrangian invariants satisfying

$d\omega/dt = d\theta/dt = 0$. Note however that vorticity is related to velocity while the passive scalar is not.

Let us now consider passive scalar turbulence. For that we add random source of fluctuations φ:

$$\partial_t \theta + (\mathbf{v} \cdot \nabla)\theta = \kappa \nabla^2 \theta + \varphi \ . \tag{1.17}$$

If the source φ produces the fluctuations of θ on some scale L then the inhomogeneous velocity field stretches, contracts and folds the field θ producing progressively smaller and smaller scales — this is the mechanism of the scalar cascade. If the rms velocity gradient is Λ then molecular diffusion is substantial at the scales less than the diffusion scale $r_d = \sqrt{\kappa/\Lambda}$. For scalar turbulence, the ratio $Pe = L/r_d$, called Peclet number, plays the role of the Reynolds number. When $Pe \gg 1$, there is an inertial interval with a constant flux of θ^2:

$$\langle(\mathbf{v}_1 \cdot \nabla_1 + \mathbf{v}_2 \cdot \nabla_2)\theta_1\theta_2\rangle = 2P \ , \tag{1.18}$$

where $P = \kappa\langle(\nabla\theta)^2\rangle = \langle\varphi\theta\rangle$ and subscripts denote the spatial points. In considering the passive scalar problem, the velocity statistics is presumed to be given. Still, the correlation function (1.18) mixes $\mathbf{v}$ and θ and does not generally allow one to make a statement on any correlation function of θ. The proper way to describe the correlation functions of the scalar at the scales much larger than the diffusion scale is to employ the Lagrangian description that is to follow fluid trajectories. Indeed, if we neglect diffusion, then the equation (1.17) can be solved along the characteristics $\mathbf{R}(t)$ which are called Lagrangian trajectories and satisfy $d\mathbf{R}/dt = \mathbf{v}(\mathbf{R}, t)$. Presuming zero initial conditions for θ at $t \to -\infty$ we write (see also Sect. 2.2.2.3 in the Gawędzki course)

$$\theta\Big(\mathbf{R}(t), t\Big) = \int_{-\infty}^{t} \varphi\Big(\mathbf{R}(t'), t'\Big) dt' \ . \tag{1.19}$$

In that way, the correlation functions of the scalar $F_n = \langle\theta(\mathbf{r}_1, t)\dots\theta(\mathbf{r}_n, t)\rangle$ can be obtained by integrating the correlation functions of the pumping along the trajectories that satisfy the final conditions $\mathbf{R}_i(t) = \mathbf{r}_i$. We consider a pumping which is Gaussian, statistically homogeneous and isotropic in space and white in time:

$$\langle\varphi(\mathbf{r}_1, t_1)\varphi(\mathbf{r}_2, t_2)\rangle = \Phi(|\mathbf{r}_1 - \mathbf{r}_2|)\delta(t_1 - t_2)$$

where the function Φ is constant at $r \ll L$ and goes to zero at $r \gg L$. The pumping provides for symmetry $\theta \to -\theta$ which makes only even correlation functions F_{2n} nonzero. The pair correlation function is as follows:

$$F_2(r,t) = \int_{-\infty}^{t} \Phi\Big(R_{12}(t')\Big)\, dt' \ . \tag{1.20}$$

Here $R_{12}(t') = |\mathbf{R}_1(t') - \mathbf{R}_2(t')|$ is the distance between two trajectories and $R_{12}(t) = r$. The function Φ essentially restricts the integration to the time interval when the distance $R_{12}(t') \le L$. Simply speaking, the stationary pair correlation function of a tracer is $\Phi(0)$ (which is twice the injection rate of θ^2) times the average time $T_2(r,L)$ that two fluid particles spent within the correlation scale of the pumping. The larger r the less time it takes for the particles to separate from r to L and the less is $F_2(r)$. Of course, $T_{12}(r,L)$ depends on the properties of the velocity field. A general theory is available only when the velocity field is spatially smooth at the scale of scalar pumping L. This so-called Batchelor regime happens, in particular, when the scalar cascade occurs at the scales less than the viscous scale of fluid turbulence (Batchelor 1959, Kraichnan 1974, Falkovich et al 2001). This requires the Schmidt number ν/κ (called Prandtl number when θ is temperature) to be large, which is the case for very viscous liquids. In this case, one can approximate the velocity difference $\mathbf{v}(\mathbf{R}_1, t) - \mathbf{v}(\mathbf{R}_2, t) \approx \hat{\sigma}(t)\mathbf{R}_{12}(t)$ with the Lagrangian strain matrix $\sigma_{ij}(t) = \nabla_j v_i$. In this regime, the distance obeys the linear differential equation

$$\dot{\mathbf{R}}_{12}(t) = \hat{\sigma}(t)\mathbf{R}_{12}(t) \ . \tag{1.21}$$

The theory of such equations is well-developed and is related to what is called Lagrangian chaos and multiplicative large deviations theory described in detail in the course of K. Gawędzki. Fluid trajectories separate exponentially as typical for systems with dynamical chaos (see, e.g. Antonsen and Ott 1991, Falkovich et al 2001): At t much larger than the correlation time of the random process $\hat{\sigma}(t)$, all moments of R_{12} grow exponentially with time and $\langle \ln[R_{12}(t)/R_{12}(0)]\rangle = \lambda t$ where λ is called a senior Lyapunov exponent of the flow (remark that for the description of the scalar we need the flow taken backwards in time which is different from that taken forward because turbulence is irreversible). Dimensionally, $\lambda = \Lambda f(Re)$ where the limit of the function f at $Re \to \infty$ is unknown. We thus obtain:

$$F_2(r) = \Phi(0)\lambda^{-1}\ln(L/r) = 2P\lambda^{-1}\ln(L/r) \ . \tag{1.22}$$

In a similar way, one shows that for $n \ll \ln(L/r)$ all F_n are expressed via F_2 and the structure functions $S_{2n} = \langle[\theta(\mathbf{r},t) - \theta(0,t)]^{2n}\rangle \simeq (P/\lambda)^n \ln^n(r/r_d)$ for $n \ll \ln(r/r_d)$. That can be generalized for an arbitrary statistics of pumping as long as it is finite-correlated in time (Balkovsky and Fouxon

1999, Falkovich et al 2001). Note that those F_{2n} anf S_{2n} are completely determined by $\Phi(0)$ which is the flux of θ^2, only sub-leading corrections depend on the fluxes of the high-order integrals.

1.4.4 Two-dimensional enstrophy cascade

Now, one can use the analogy between passive scalar and vorticity in 2d (Kraichnan 1967,Falkovich and Lebedev 1994). For the enstrophy cascade, one derives the flux relation analogous to (1.18):

$$\langle (\mathbf{v}_1 \cdot \nabla_1 + \mathbf{v}_2 \cdot \nabla_2)\omega_1\omega_2 \rangle = 2D \ , \qquad (1.23)$$

where $D = \langle \nu(\nabla\omega)^2 \rangle$. The flux relation along with $\omega = curl\,\mathbf{v}$ suggests the scaling $\delta v(r) \propto r$ that is velocity being close to spatially smooth (of course, it cannot be perfectly smooth to provide for a nonzero vorticity dissipation in the inviscid limit, but the possible singularitites are indeed shown to be no stronger than logarithmic). That makes the vorticity cascade similar to the Batchelor regime of passive scalar cascade with a notable change in that the rate of stretching λ acting on a given scale is not a constant but is logarithmically growing when the scale decreases. Physically, for smaller blobs of vorticity there are more large-scale velocity gradients that are able to stretch them. Since λ scales as vorticity, the law of renormalization can be established from dimensional reasoning and one gets $\langle \omega(\mathbf{r},t)\omega(0,t) \rangle \sim [D\ln(L/r)]^{2/3}$ which corresponds to the energy spectrum $E_k \propto D^{2/3}k^{-3}\ln^{-1/3}(kL)$. High-order correlation functions of vorticity are also logarithmic, for instance, $\langle \omega^n(\mathbf{r},t)\omega^n(0,t) \rangle \sim [D\ln(L/r)]^{2n/3}$. Note that both passive scalar in the Batchelor regime and vorticity cascade in 2d are universal that is determined by the single flux (P and D respectively) despite the existence of high-order conserved quantities. Experimental data and numeric simulations support those conclusions (Falkovich et al 2001, Tabeling 2002).

It is instructive to compare the flux relations (1.5, 1.10, 1.14, 1.18, 1.23) with the mass flux relation described in Section 3.7.3 of Cardy's course.

1.5 Zero modes and anomalous scaling

How one builds the Lagrangian description when the velocity is not spatially smooth, for example, that of the energy cascades in the inertial interval? Again, the only exact relation one can derive for two fluid particles separated by a distance in the inertial interval is for the Lagrangian time derivative of the squared velocity difference (Falkovich et al 2001):

$$\left\langle \frac{d|\delta \mathbf{v}|^2}{dt} \right\rangle = 2\epsilon$$

— this is the Lagrangian counterpart to (1.5,1.10,1.14,1.29). One can *assume* that the statistics of the distances between particles is also determined by the energy flux. That assumption leads, in particular, to the Richardson law for the asymptotic growth of the inter-particle distance:

$$\langle R_{12}^2(t) \rangle \sim \epsilon t^3 \, , \tag{1.24}$$

first inferred from atmospheric observations (in 1926) and later from experimental data on the energy cascades both in 3d and in 2d. There is no consistent theoretical derivation of (1.24) and it is unclear whether it is exact (likely to be in 2d) or just approximate (possible in 3d). Semi-heuristic argument usually presented in textbooks is based on the mean-field estimate: $\dot{\mathbf{R}}_{12} = \delta\mathbf{v}(\mathbf{R}_{12}, t) \sim (\epsilon R_{12})^{1/3}$ which upon integration gives: $R_{12}^{2/3}(t) - R_{12}^{2/3}(0) \sim \epsilon^{1/3}t$. While this argument is at best a crude estimate in 3d (where there is no definite velocity scaling since every moment has its own exponent ζ_n) we use it to discuss implications for the passive scalar†.

For two trajectories, the Richardson law gives the separation time from r to L: $T_2(r, L) \sim \epsilon^{-1/3}[L^{2/3} - r^{2/3}]$. Note that $T_2(r, L)$ has a finite limit at $r \to 0$ — infinitesimally close trajectories separate in a finite time. That leads to non-uniqueness of Lagrangian trajectories (non-smoothness of the velocity field means that the equation $\dot{\mathbf{R}} = \mathbf{v}(\mathbf{R})$ is non-Lipschitz). As discussed in much details elsewhere (see Falkovich et al 2001 and Lecture 4 of the Gawędzki course), that leads to a finite dissipation of a transported passive scalar even without any molecular diffusion (which corresponds to a dissipative anomaly and time irreversibility). Indeed, substituting $T_2(r, L)$ into (1.20), one gets the steady-state pair correlation function of the passive scalar: $F_2(r) \sim \Phi(0)\epsilon^{-1/3}[L^{2/3} - r^{2/3}]$ as suggested by Oboukhov (1949) and Corrsin (1952). The structure function is then $S_2(r) \sim \Phi(0)\epsilon^{-1/3}r^{2/3}$. Experiments measuring the scaling exponents $\sigma_n = d\ln S_n(r)/d\ln r$ generally give σ_2 close to 2/3 but higher exponents deviating from the straight line even stronger than the exponents of the velocity in 3d as seen in Fig. 1.4. Moreover, the scalar exponents σ_n are anomalous even when advecting velocity has a normal scaling like in 2d energy cascade (to be described in Sec. 1.6 below).

---

† What matters here and below is that in a non-smooth flow $R_{12}^a(t) - R_{12}^a(0) \sim t$ with $a < 1$, not the precise value of a

To explain the dependence $\sigma(n)$ and describe multi-point correlation functions or high-order structure functions one needs to study multi-particle statistics. Here an important question is what memory of the initial configuration remains when final distances far exceed initial ones. To answer this question one must analyze the conservation laws of turbulent diffusion. We now describe a general concept of conservation laws which, while conserved only on the average, still determine the statistical properties of strongly fluctuating systems. In a random system, it is always possible to find some fluctuating quantities which ensemble averages do not change. We now ask a more subtle question: is it possible to find quantities that are expected to change on the dimensional grounds but they stay constant (Falkovich et al 2001, Falkovich and Sreenivasan 2006). Let us characterize n fluid particles in a random flow by inter-particle distances R_{ij} (between particles i and j) as in Figure 1.5. Consider homogeneous functions f of inter-particle distances with a nonzero degree ζ, i.e. $f(\lambda R_{ij}) = \lambda^\zeta f(R_{ij})$. When all the distances grow on the average, say according to $< R_{ij}^2 >\propto t^a$, then one expects that a generic function grows as $f \propto t^{a\zeta/2}$. How to build (specific) functions that are conserved on the average, and which ζ-s they have? As the particles move in a random flow, the n-particle cloud grows in size and the fluctuations in the shape of the cloud decrease in magnitude. Therefore, one may look for suitable functions of size and shape that are conserved because the growth of distances is compensated by the decrease of shape fluctuations.

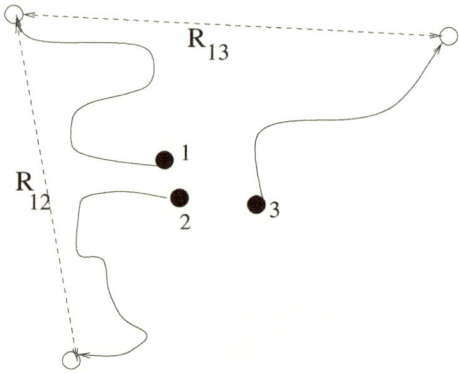

Fig. 1.5. Three fluid particles in a flow.

For the simplest case of Brownian random walk, inter-particle distances grow by the diffusion law: $\langle R_{ij}^2(t)\rangle = R_{ij}^2(0) + \kappa t$, $\langle R_{ij}^4(t)\rangle = R_{ij}^4(0) + 2(d + 2)\left[R_{ij}^2(0)\kappa t + \kappa^2 t^2\right]/d$, etc. Here d is the space dimensionality. Two particles are characterized by a single distance. Any positive power of this distance grows on the average. For many particles, one can build conserved quantities by taking the differences where all powers of t cancel out: $f_2 = \langle R_{12}^2 - R_{34}^2 \rangle$, $f_4 = \langle 2(d+2)R_{12}^2 R_{34}^2 - d(R_{12}^4 + R_{34}^4) \rangle$, etc. These polynomials are called harmonic since they are zero modes of the Laplacian in the $2d$-dimensional space of $\mathbf{R}_{12}$, $\mathbf{R}_{13}$. One can write the Laplacian as $\Delta = R^{1-2d}\partial_R R^{2d-1}\partial_R + \Delta_\theta$, where $R^2 = R_{12}^2 + R_{13}^2$ and Δ_θ is the angular Laplacian on $2d-1$-dimensional unit sphere. Introducing the angle, $\theta = \arcsin(R_{12}/R)$, which characterizes the shape of the triangle, we see that the conservation of both $f_2 = \langle R^2 \cos 2\theta \rangle$ and $f_4 = \langle R^4[(d+1)\cos^2 2\theta - 1] \rangle$ can be also described as due to cancelation between the growth of the radial part (as powers of t) and the decay of the angular part (as inverse powers of t). For n particles, the polynomial that involves all distances is proportional to R^{2n} (i.e. $\zeta_n = n$) and the respective shape fluctuations decay as t^{-n}.

The scaling exponents of the zero modes are thus determined by the laws that govern decrease of shape fluctuations. The zero modes, which are conserved statistically, exist for turbulent macroscopic diffusion as well. However, there is a major difference since the velocities of different particles are correlated in turbulence. Those mutual correlations make shape fluctuations decaying slower than t^{-n} so that the exponents of the zero modes, ζ_n, grow with n slower than linearly. This is very much like the total energy of the cloud of attracting particles does not grow linearly with the number of particles. Indeed, power-law correlations of the velocity field lead to super-diffusive behavior of inter-particle separations: the farther particles are, the faster they tend to move away from each other, as in Richardson's law of diffusion. That is the system behaves as if there was an attraction between particles that weakens with the distance, though, of course, there is no physical interaction among particles (but only mutual correlations because they are inside the correlation radius of the velocity field). Let us stress that while zero modes of multi-particle evolution exist for all velocity fields—from those that are smooth to those that are extremely rough as in Brownian motion—only those non-smooth velocity fields with power-law correlations provide for an anomalous scaling. Zero modes were discovered in Gawedzki and Kupiainen 1995, Shraiman and Siggia 1995, Chertkov et al 1995 and then described in Chertkov and Falkovich 1996, Bernard et al 1996, Balkovsky and Lebedev 1998.

The existence of multi-particle conservation laws indicates the presence of a long-time memory and is a reflection of the coupling among the particles due to the simple fact that they are all in the same velocity field.

We now ask: How does the existence of these statistical conservation laws (called martingales in the probability theory) lead to anomalous scaling of fields advected by turbulence? According to (1.19), the correlation functions of θ are proportional to the times spent by the particles within the correlation scales of the pumping. The structure functions of θ are differences of correlation functions with different initial particle configurations as, for instance, $S_3(r_{12}) \equiv \langle [\theta(\mathbf{r}_1) - \theta(\mathbf{r}_2)]^3 \rangle = 3\langle \theta^2(\mathbf{r}_1)\theta(\mathbf{r}_2) - \theta(\mathbf{r}_1)\theta^2(\mathbf{r}_2) \rangle$. In calculating S_3, we are thus comparing two histories: the first one with two particles initially close to the position $\mathbf{r}_1$ and one particle at $\mathbf{r}_2$, and the second one with one particle at $\mathbf{r}_1$ and two particles at $\mathbf{r}_2$— see Fig 1.6. That is, S_3 is proportional to the time during which one can distinguish

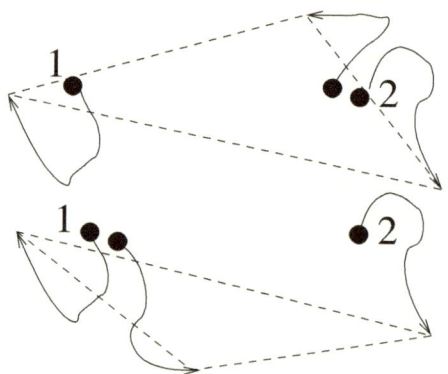

Fig. 1.6. Two configurations (upper and lower) whose difference determines the third structure function.

one history from another, or to the time needed for an elongated triangle to relax to the equilateral shape. That time grows with r_{12} (as it takes longer to forget more elongated triangle) by the law that can be inferred from the law of the decrease of the shape fluctuations of a triangle.

Quantitative details can be worked out for the white in time velocity (Kraichnan 1968). Profound insight of Kraichnan was that it is spatial rather than temporal non-smoothness of the velocity that is crucial for an anomalous scaling. The Kraichnan model is described in much detail in the course by Gawędzki, here we mention few salient points. The velocity ensemble is defined by the second moment:

$$\langle v^i(\mathbf{r}, t)v^j(0,0) \rangle = \delta(t)\left[D_0\delta_{ij} - d_{ij}(\mathbf{r}) \right],$$

$$d_{ij} = D_1 r^\xi \left[(d - 1 + \xi)\delta^{ij} - \xi r^i r^j r^{-2} \right]. \tag{1.25}$$

Here the exponent $\xi \in [0, 2]$ is a measure of the velocity non-smoothness with $\xi = 2$ corresponding to a smooth velocity while $\xi = 0$ to a velocity very rough in space (distributional). Richardson-Kolmogorov scaling of the energy cascade corresponds to $\xi = 4/3$. Lagrangian flow is a Markov random process for the Kraichnan ensemble (1.25). Every fluid particle undergoes a Brownian random walk with the so-called eddy diffusivity D_0. The PDF $P(r, t)$ for two particles to be separated by r after time t satisfies the diffusion equation (see e.g. Falkovich et al 2001)

$$\partial_t P = L_2 P, \quad L_2 = d_{ij}(\mathbf{r})\nabla^i \nabla^j = D_1(d-1)r^{1-d}\partial_r r^{d+\xi-1}\partial_r, \quad (1.26)$$

with the scale-dependent diffusivity $D_1(d-1)r^\xi$. The asymptotic solution of (1.26) is $P(r, t) = r^{d-1}t^{d/(2-\xi)}\exp(-\text{const } r^{2-\xi}/t)$, log-normal for $\xi = 2$. For $\xi = 4/3$, it reproduces, in particular, the Richardson law. Multi-particle probability distributions also satisfy diffusion equations in the Kraichnan model as well as all the correlation functions of θ. Multiplying (1.16) by $\theta_2 \dots \theta_{2n}$ and averaging over the Gaussian statistics of $\mathbf{v}$ and φ one derives

$$\partial_t F_{2n} = L_{2n}F_{2n} + \sum_{l,m} F_{2n-2}\Phi(\mathbf{r}_{lm}), \quad L_{2n} = \sum d_{ij}(\mathbf{r}_{lm})\nabla^i_l \nabla^j_m. \quad (1.27)$$

This equation enables one, in principle, to derive inductively all steady-state F_{2n} starting from F_2. The equation $\partial_t F_2(r, t) = L_2 F_2(r, t) + \Phi(r)$ has a steady solution $F_2(r) = 2[\Phi(0)/(2-\xi)d(d-1)D_1][dL^{2-\xi}/(d-2+\xi) - r^{2-\xi}]$, which has the Corrsin-Oboukhov form for $\xi = 4/3$. Further, F_4 contains the so-called forced solution having the normal scaling $2(2 - \xi)$ but also, remarkably, a zero mode Z_4 of the operator L_4: $L_4 Z_4 = 0$. Such zero modes necessarily appear (to satisfy the boundary conditions at $r \simeq L$) for all $n > 1$ and the scaling exponents of Z_{2n} are generally different from $n\gamma$ that is anomalous. In calculating the scalar structure functions, all terms cancel out except a single zero mode (called irreducible because it involves all distances between $2n$ points). Analytically and numerical calculations of Z_n and their scaling exponents σ_n (described in detail in the course of K. Gawędzki and in the review Falkovich et al 2001) give σ_n lying on a convex curve (see Fig. 1.4) which saturates (Balkovsky and Lebedev 1998) to a constant at large n. Such saturation is a signature that most singular structures in a scalar field are shocks like in Burgers turbulence, the value σ_n at $n \to \infty$ is the fractal codimension of fronts in space (Celani et al 2001).

The existence of statistical conserved quantities breaks the scale invariance of scalar statistics in the inertial interval and explains why scalar turbulence knows about pumping "more" than just the value of the flux. Here again the statistics in the inertial interval, apart from the flux of θ^2, depends on the infinity of pumping-related parameters. However, those parameters neither are fluxes of θ^n, nor we can interpret them as any other fluxes. At the present level of understanding, we thus describe an anomalous scaling in Burgers and in passive scalar in quite different terms. Of course, the qualitative appeal to structures (shocks) is similar but the nature of the conservation laws is different. The anomalies produced by dynamically conserved quantities (like anomalous scaling in Burgers and time irreversibility in all cases of turbulence) are qualitatively different from the anomalies produced by statistically conserved quantities (like breakdown of scale invariance in passive scalar turbulence). Indeed, dissipation is a singular perturbation which breaks conservation of dynamical integrals of motion and imposes (one or many) flux-constancy conditions, very much similar to quantum anomalies. On the contrary, there are no cascades of conserved quantity related to zero modes, nor their conservation is broken by dissipation. Anomalous scaling of zero modes is due to correlations between different fluid trajectories. On the other hand, the two types of anomalies are related intimately: the flux constancy requires a certain degree of velocity non-smoothness, which generally leads to an anomalous scaling of zero modes.

Both symmetries, one broken by pumping (scale invariance) and another by damping (time reversibility) are not restored even when $r/L \to 0$ and $r_d/r \to 0$.

For the vector field (like velocity or magnetic field in magnetohydrodynamics) the Lagrangian statistical integrals of motion may involve both the coordinate of the fluid particle and the vector it carries. Such integrals of motion were built explicitly and related to the anomalous scaling for the passively advected magnetic field in the Kraichnan ensemble of velocities (Falkovich et al 2001). Doing that for velocity that satisfies the 3d Navier-Stokes equation remains a task for the future.

1.6 Inverse cascades

Here we consider inverse cascades and discover that, while time reversibility remains broken, the scale invariance is restored in the inertial interval. Moreover, even wider symmetry of conformal invariance may appear there.

1.6.1 Passive scalar in a compressible flow

Similar to (1.20) one can derive from (1.19)

$$\langle \theta(t, \mathbf{r}_1) \ldots \theta(t, \mathbf{r}_{2n}) \rangle = \int_0^t dt_1 \ldots dt_n$$
$$\times \langle \Phi(R(t_1|T, \mathbf{r}_{12})) \ldots \Phi(R(t_n|T, \mathbf{r}_{2n-1,2n})) \rangle + \ldots , \qquad (1.28)$$

The functions Φ in (1.28) restrict integration to the time intervals where $R_{ij} < L$. If the Lagrangian trajectories separate, the correlation functions reach at long times the stationary form for all r_{ij}. Such steady states correspond to a direct cascade of the tracer (i.e. from large to small scales) considered above. That generally takes place in incompressible and weakly compressible flows.

It is intuitively clear that in compressible flows the regions of compressions can trap fluid particles counteracting their tendency to separate. Indeed, one can show that particles cluster in flows with high enough compressibility (Chertkov et al 1998, Gawędzki and Vergassola 2000). In particular, the solution of the Problem 3 shows that all the Lyapunov exponents are negative when the compressibility degree of a short-correlated flow exceeds $d/4$ (see (1.45) and (2.53) of Gawedzki's course below). Even in the non-smooth flow with high enough compressibility, the trajectories are unique, particles that start from the same point will remain together throughout the evolution (Gawędzki and Vergassola 2000). That means that advection preserves all the single-point moments $\langle \theta^N \rangle(t)$. Note that the conservation laws are statistical: the moments are not dynamically conserved in every realization, but their average over the velocity ensemble are. In the presence of pumping, the moments are the same as for the equation $\partial_t \theta = \varphi$ in the limit $\kappa \to 0$ (nonsingular now). It follows that the single-point statistics is Gaussian, with $\langle \theta^2 \rangle$ coinciding with the total injection $\Phi(0)t$ by the forcing. That growth is produced by the flux of scalar variance toward the large scales. In other words, the correlation functions acquire parts which are independent of r and grow proportional to time: when Lagrangian particles cluster rather than separate, tracer fluctuations grow at larger and larger scales — phenomenon that can be loosely called an inverse cascade of a passive tracer (Chertkov et al 1998, Gawędzki and Vergassola 2000). As is clear from (1.28), correlation functions at very large scales are related to the probability for initially distant particles to come close. In a strongly compressible flow, the trajectories are typically contracting, the particles tend to approach and the distances will reduce to the forcing correlation length L (and smaller) for long enough times. On a particle language, the larger the time the large the distance starting from which particle come within L. The correlations of the field θ at larger and larger scales are therefore established as time increases, signaling the inverse cascade process.

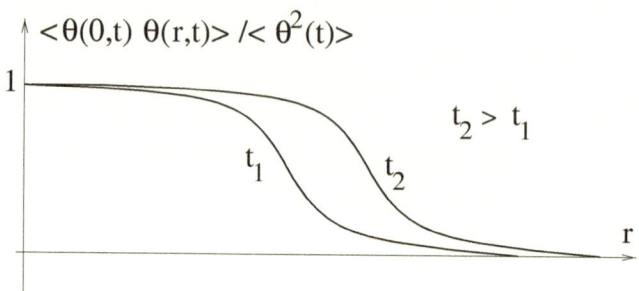

Fig. 1.7. Growth of large-scale correlations with time.

The uniqueness of the trajectories greatly simplifies the analysis of the PDF $\mathcal{P}(\delta\theta, r)$. Indeed, the structure functions involve initial configurations with just two groups of particles separated by a distance r. The particles explosively separate in the incompressible case and we are immediately back to the full N-particle problem. Conversely, the particles that are initially in the same group remain together if the trajectories are unique. The only relevant degrees of freedom are then given by the intergroup separation and we are reduced to a two-particle dynamics. It is therefore not surprising that the statistics of the passive tracer is scale invariant in the inverse cascade regime (Gawędzki and Vergassola 2000).

An example of strongly compressible flow is given by Burgers turbulence (1.8) where there is clustering (in shocks) for the majority of trajectories (full measure in the inviscid limit). Considering passive scalar in such a flow, $\theta_t + u\theta_x - \kappa\Delta\theta = \phi$, we conclude that it undergoes an inverse cascade. The statistics of θ is scale invariant at the scales exceeding the correlation scale of the pumping ϕ. While the limit $\kappa \to 0$ is regular (i.e. no dissipative anomaly), the statistics is time irreversible because of the flux towards large scales. It is instructive to compare u and θ which are both Lagrangian invariants (tracers) in the unforced undamped limit. Yet passive quantity θ (and all its powers) go to large scales under pumping while all powers of u cascade towards small scales and are absorbed by viscosity. Physically, the difference is evidently due to the fact that the trajectory evidently depends on the value of u it carries, the larger the velocity the faster it ends in a shock and dissipates the energy and other integrals. Formally, for active tracers like u^n one cannot write a formula like (1.28) obtained by two independent averages over the force and over the trajectories.

1.6.2 Inverse energy cascade in two dimensions

For the inverse energy cascade, there is no consistent theory except for the flux relation that can be derived similarly to (1.14):

$$S_3(r) = 4\epsilon r/3 . \qquad (1.29)$$

This scaling one can also get from phenomenological dimensional arguments, though in two seemingly unrelated ways. Consider the velocity difference v_r at the distance r. On the one hand, one may require that the kinetic energy v_r^2 divided by the typical time r/v_r must be constant and equal to the energy flux, ϵ: $v_r^3 \sim \epsilon r$. On the other hand, it can be argued that vorticity, which cascades to small scales, must be in equipartition in the inverse cascade range. If this is the case, the enstrophy $r^d \omega_r^2$ accumulated in a volume of size r is proportional to the typical time r/v_r at such scale, i.e. $r^d \omega_r^2 \sim r/v_r$. Using $\omega_r \sim v_r/r$ we derive $v_r^3 \sim r^{3-d}$ which for $d = 2$ is exactly the requirement of constant energy flux. Amazingly, the requirements of vorticity equipartition (i.e. equilibrium) and energy flux (i.e. turbulence) give the same Kolmogorov-Kraichnan scaling in 2d. Let us stress that (1.29) means that time reversibility is broken in the inverse cascade. Experiments (Tabeling 2002, Kellay and Goldburg 2002, Chen et al 2006) and numerical simulations (Boffetta et al 2000), however, demonstrate a scale-invariant statistics with the vorticity having scaling dimension $2/3$: $\omega_r \propto r^{-2/3}$. In particular, $S_2 \propto r^{2/3}$ which corresponds to $E_k \propto k^{-5/3}$. It is ironic that probably the most widely known statement on turbulence, the 5/3 spectrum suggested by Kolmogorov for 3d, is not correct in this case (even though the true scaling is close) while it is probably exact in the Kraichnan's inverse 2d cascade. Qualitatively, it is likely that the absence of anomalous scaling in the inverse cascade is associated with the growth of the typical turnover time (estimated, say, as $r/\sqrt{S_2}$) with the scale. As the inverse cascade proceeds, the fluctuations have enough time to get smoothed out as opposite to the direct cascade in 3d, where the turnover time decreases in the direction of the cascade.

Remarkably, there are indications that scale invariance can be extended to conformal invariance at least for some properties of 2d turbulence (Bernard et al 2006). Under conformal transformations the lengths are re-scaled non-uniformly yet the angles between vectors are left unchanged (a useful property in navigation cartography where it is often more important to aim in the right direction than to know the distance). Conformal invariance has been discovered by analyzing the large-scale statistics of the boundaries of vorticity clusters, i.e. large-scale zero-vorticity (nodal) lines. In equilibrium critical phenomena, cluster boundaries in the continuous limit of vanishingly small lattice size were recently found to belong to a remarkable class of curves that can be mapped into Brownian walk. That class is called Schramm-Loewner Evolution or SLE curves (Schramm 2000, Gruzberg and Kadanoff 2004, Lawler 2005, Cardy 2005, Bauer and Bernard 2006). Namely, consider a curve $\gamma(t)$ that starts at a point on the boundary of the half-plane H (by conformal invariance any planar domain is equivalent to the upper half plane). One can map the half-plane H minus the curve $\gamma(t)$ back onto H by an analytic function $g_t(z)$ which is unique upon imposing the condition $g_t(z) \sim z + 2t/z + O(1/z^2)$ at infinity. The growing tip of the curve is mapped into a real point $\xi(t)$. Loewner found in 1923 that the conformal map $g_t(z)$ and the curve $\gamma(t)$ are fully parametrized by the driving function $\xi(t)$. Almost eighty years later, Schramm (2000) considered random curves in planar domains and showed that their statistics is conformal invariant if $\xi(t)$ is a Brownian walk, i.e. its increments are identically and independently distributed and $\langle(\xi(t) - \xi(0))^2\rangle = \kappa t$. In simple words, the locality in time of the Brownian walk translates into the local scale-invariance of SLE curves, i.e. conformal invariance. SLE_κ provide a natural classification (by the value of the diffusivity κ) of boundaries of clusters of 2d critical phenomena described by conformal field theories (see Gruzberg and Kadanoff 2004, Lawler 2005, Cardy 2005, Bauer and Bernard 2006 for a review).

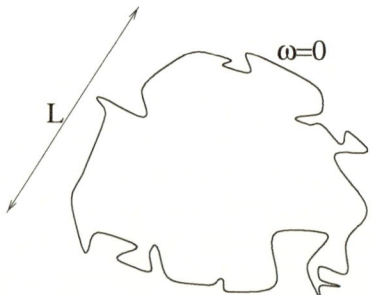

Fig. 1.8. Vorticity nodal line with the gyration radius L.

The fractal dimension of SLE_κ curves is known to be $D_\kappa = 1 + \kappa/8$ for $\kappa < 8$. To establish possible link between turbulence and critical phenomena, let us try to relate the Kolmogorov-Kraichnan phenomenology to the fractal dimension of the boundaries of vorticity clusters. Note that one ought to distinguish between the dimensionality 2 of the full vorticity level set (which is space-filling) and a single zero-vorticity line that encloses a large-scale cluster. Consider the vorticity cluster of gyration radius L which has the "outer boundary" of perimeter P (that boundary is the part of the zero-vorticity line accessible from outside, see Fig. 1.8 for an illustration). The vorticity flux through the cluster, $\int \omega dS \sim \omega_L L^2$, must be equal to the velocity circulation along the boundary, $\Gamma = \oint \mathbf{v} \cdot d\ell$. The Kolmogorov-Kraichnan scaling is $\omega_L \sim \epsilon^{1/3} L^{-2/3}$ (coarse-grained vorticity decreases with scale because contributions with opposite signs partially cancel) so that the flux is $\propto L^{4/3}$. As for circulation, since the boundary turns every time it meets a vortex, such a contour is irregular on scales larger than the pumping scale. Therefore, only the velocity at the pumping scale L_f is expected to contribute to the circulation, such velocity can be estimated as $(\epsilon L_f)^{1/3}$ and it is independent of L. Hence, circulation should be proportional to the perimeter, $\Gamma \propto P$, which gives $P \propto L^{4/3}$, i.e. the fractal dimension of the exterior of the vorticity cluster is expected to be $4/3$. This remarkable dimension corresponds to a self-avoiding random walk (SLE curve) which is also known to be an exterior boundary (without self-intersections) of percolation cluster (yet another SLE curve). Data analysis of the zero-vorticity lines have shown that indeed within an experimental accuracy their statistics is indistinguishable from percolation clusters while that of their exterior boundary from the statistics of self-avoiding random walk (Bernard et al 2006). Whether the statistics of the zero-vorticity isolines indeed falls into the simplest universality class of critical phenomena (that of percolation) deserves more study.

Let us briefly discuss wave turbulence from the viewpoint of conformal invariance. Gaussian scalar field in 2d is conformal invariant if its correlation function is logarithmic i.e. the spectral density decays as k^{-2}. Such is the case, for instance, for the fluid height in gravitational-capillary weak wave turbulence on a shallow water (see Zakharov et al 1992, Sect. 5.1.2). It is interesting if deviations from Gaussianity due to wave interaction destroy conformal invariance. Another interesting example is the inverse cascade of 2d strong optical turbulence described by the Nonlinear Schrödinger Equation. Numerics hint that in the case of a stable growing condensate, the statistics of the finite-scale fluctuations approach Gaussian with a logarithmic correlation function (Dyachenko and Falkovich 1996).

1.7 Conclusion

We reiterate the conclusions on the status of symmetries in turbulence. Turbulence statistics is always time-irreversible.

Weak turbulence is scale invariant and universal (determined solely by the flux value). It is generally not conformal invariant.

Strong turbulence: Direct cascades often have symmetries broken by pumping (scale invariance, isotropy) non-restored in the inertial interval. In other words, statistics at however small scales is sensitive to other characteristics of pumping besides the flux. That can be alternatively explained in terms of either structures or statistical conservation laws (zero modes). Anomalous scaling in a direct cascade may well be a general rule apart from some degenerate cases like passive scalar in the Batchelor case (where all the zero modes have the same scaling exponent, zero, as the pair correlation function). Inverse cascades in systems with strong interaction may be not only scale invariant but also conformal invariant. It is an example of emerging or restored symmetry.

For Lagrangian invariants, we explain the difference between direct and inverse cascades in terms of separation or clustering of fluid particles. Generally, it seems natural that the statistics within the pumping correlation scale (direct cascade) is more sensitive to the details of the pumping statistics than the statistics at much larger scales (inverse cascade).

1.8 Exercises

1.8.1 Problems

(i) Show that $n_k \propto k^{-m-d}$ turns the three-wave collision integral $I_k^{(3)}$ into zero. Show that it corresponds to a constant energy flux which sign is given by the derivative of the collision integral with respect to the exponent of the spectrum.

(ii) A general equilibrium solution of $I_k^{(3)} = 0$ depends on the energy and the momentum of the wave system: $n(\mathbf{k}, T, \mathbf{u}) = T[\omega_k - (\mathbf{k} \cdot \mathbf{u})]^{-1}$ (Doppler-shifted Rayleigh-Jeans distribution). A general non-equilibrium solution depends on the fluxes P and $\mathbf{R}$ of the energy and momentum respectively. Find the form of the weakly anisotropic correction to the isotropic turbulence spectrum.

(iii) For the two-particle distance evolving in a spatially smooth random flow according to (1.21), $\dot{\mathbf{R}}(t) = \hat{\sigma}(t)\mathbf{R}(t)$, consider the Jacobi matrix defined by $R_i(t) = W_{ij}(t)R_j(t)$ — see also Sects. 2.2.2.3 and 2.3.2 in the Gawędzki course. An initial infinitesimal sphere evolves into an elongated ellipsoid with the inertia tensor $I(t) = W(t)W^T(t)$. Lyapunov exponents are the asymptotic in time eigenvalues of $W^T(t)W(t)$ which stabilizes in every realization. They can be also found from the tensor I, which on the contrary, rotates all the time. Decompose $I = O^T \Lambda O$ where O is an orthogonal matrix composed of the eigenvectors of I and Λ is a diagonal matrix with the eigenvalues $e^{2\rho_1}, \ldots, e^{2\rho_d}$. Derive the equation for $I(t)$ and $\rho_i(t)$ and find the Lyapunov exponents as $\lambda_i = \lim_{t\to\infty} \rho_i(t)/t$ for the short-correlated isotropic Gaussian strain $\langle \sigma_{ij}(0)\sigma_{kl}(t) \rangle = C_{ijkl}\delta(t)$.

1.8.2 Solutions

(i) **Exercise 1.1** We write the collision integral as follows

$$I_k^{(3)} = \int \left(U_{k12} - U_{1k2} - U_{2k1} \right) dk_1 dk_2$$

with

$$U_{123} = \pi \left[n_2 n_3 - n_1 (n_2 + n_3) \right] |V_{123}|^2 \delta \left(\mathbf{k}_1 - \mathbf{k}_2 - \mathbf{k}_3 \right) \delta \left(\omega_1 - \omega_2 - \omega_3 \right).$$

Here and below i stands for $\mathbf{k}_i$. To evaluate $I_k^{(3)}$ on $n_k \propto k^{-m-d}$ we first integrate over the directions of $\mathbf{k}_i$. In an isotropic medium, the interaction coefficient V_{k12} depends only on the scalar products of $\mathbf{k}$ and $\mathbf{k}_i$. Using the additional condition $\mathbf{k} - \mathbf{k}_1 - \mathbf{k}_2 = 0$ one can express V_{k12}, like the frequency $\omega(k)$, as a function of wave numbers. Then, only the δ-function of wave vectors is to be integrated over the angles in the $\mathbf{k}_1, \mathbf{k}_2$ space. The result of the angular integration is non-zero only if one can form a triangle out of $\mathbf{k}$, $\mathbf{k}_1$ and $\mathbf{k}_2$. Denote $\Theta(k_1, k_2, k_3)$ the product of step functions $\theta(k_1 + k_2 - k)\theta(k + k_2 - k_1)\theta(k + k_1 - k_2)$ that ensures the triangular inequalities $k < k_1 + k_2$, $k_1 < k + k_2$ and $k_2 < k + k_1$. We introduce $\hat{k}_i = \mathbf{k}_i / k_i$ and $dk_i = k_i^{d-1} dk_i d\hat{k}_i$. Applying the formula $\int \delta \left(\mathbf{f} - g\hat{g} \right) d\hat{g} = 2\delta(f^2 - g^2)/g^{d-2}$ to the integration over $\hat{k}_2$ we obtain

$$H_d \equiv \int \delta(\mathbf{k} - \mathbf{k}_1 - \mathbf{k}_2) d\hat{k}_1 d\hat{k}_2 = \int d\hat{k}_1 \, 2 k_2^{2-d} \delta \left(k^2 + k_1^2 - k_2^2 - 2\mathbf{k} \cdot \mathbf{k}_1 \right).$$

We confine ourselves to physical dimensions $d = 2$ and $d = 3$ though one can easily generalize the following to arbitrary d. In $d = 2$ we have

$$H_2 = 4 \int_0^\pi \delta \left(k^2 + k_1^2 - k_2^2 - 2kk_1 \cos \theta \right) d\theta = \frac{2\Theta(k, k_1, k_2)}{kk_1 \sin \theta^0} = \frac{\Theta(k, k_1, k_2)}{\Delta(k, k_1, k_2)},$$

where $\cos \theta^0 = (k^2 + k_1^2 - k_2^2)/(2kk_1)$ and $\Delta(k, k_1, k_2) = (1/2) \left[2 \left(k^2 k_1^2 + k^2 k_2^2 + k_1^2 k_2^2 \right) - k^4 - k_1^4 - k_2^4 \right]^{1/2}$ is the area of the triangle formed by the vectors $\mathbf{k}, \mathbf{k}_1$ and $\mathbf{k}_2$. We have used the fact that the triangular inequalities are equivalent to one condition $|(k^2 + k_1^2 - k_2^2)/(2kk_1)| \leq 1$. Analogous calculation in $d = 3$ produces

$$H_3 = \frac{4\pi}{k_2} \int_0^\pi \delta \left(k^2 + k_1^2 - k_2^2 - 2kk_1 \cos \theta \right) \sin \theta \, d\theta = \frac{2\pi \Theta(k, k_1, k_2)}{kk_1 k_2}.$$

Observe that H_d depends only on the wave numbers k, k_1 and k_2, invariant with respect to their permutations and is non-zero only if triangular inequalities are satisfied:

$$I_k^{(3)} = \int_0^\infty k_1^{d-1} dk_1 \int_0^\infty k_2^{d-1} dk_2 H(k, k_1, k_2) \left(\tilde{U}_{k12} - \tilde{U}_{1k2} - \tilde{U}_{2k1} \right) ,$$

$$(1.30)$$

$$\tilde{U}_{123} = \pi \left[n_2 n_3 - n_1 (n_2 + n_3) \right] |V_{123}|^2 \delta (\omega_1 - \omega_2 - \omega_3) .$$

We change the integration variables from k_1, k_2 to t $\omega(k_1) = \omega_1$, $\omega(k_2) = \omega_2$. Using $\omega(k) \propto k^\alpha$ we obtain for $\pi(2k)^{d-1} I(k)/v(k) \equiv I(\omega)$

$$I(\omega) = \int_0^\infty \int_0^\infty d\omega_1 d\omega_2 \left[R(\omega, \omega_1, \omega_2) - R(\omega_1, \omega, \omega_2) - R(\omega_2, \omega, \omega_1) \right] \quad (1.31)$$

$$= \int_0^\omega Q(\omega, \omega_1, \omega - \omega_1)[n(\omega_1)n(\omega - \omega_1) - n(\omega)(n(\omega_1) + n(\omega - \omega_1))]d\omega_1$$

$$-2 \int_\omega^\infty Q(\omega_1, \omega, \omega_1 - \omega)[n(\omega)n(\omega_1 - \omega) - n(\omega_1)(n(\omega) + n(\omega_1 - \omega))] \, d\omega_1.$$

$$R(\omega, \omega_1, \omega_2) = C|V(\omega, \omega_1, \omega_2)|^2 H_d(\omega, \omega_1, \omega_2)(\omega\omega_1\omega_2)^{-1+d/\alpha}\delta(\omega - \omega_1 - \omega_2)$$

$$\times [n_1 n_2 - n_\omega(n_1 + n_2)] \equiv Q(\omega, \omega_1, \omega_2)\delta(\omega - \omega_1 - \omega_2)[n_1 n_2 - n_\omega(n_1 + n_2)].$$

Here $v(k) = d\omega(k)/dk$. Homogeneity properties

$$V(\lambda\mathbf{k}, \lambda\mathbf{k}_1, \lambda\mathbf{k}_2) = \lambda^m V(\mathbf{k}, \mathbf{k}_1, \mathbf{k}_2), \quad H_d(\lambda\mathbf{k}, \lambda\mathbf{k}_1, \lambda\mathbf{k}_2) = \lambda^{-d} H_d(k, k_1, k_2)$$

allow us to write $Q(\omega, \omega_1, \omega - \omega_1) = \omega^\gamma f(\omega/\omega_1)$, where $\gamma = 2(m + d)/\alpha - 3$. All the information about the interactions is contained in one number γ and in one function $f(x)$. One solution that turns the collision integral into zero is the equilibrium Rayleigh-Jeans distribution ,$n(\omega) \propto \omega^{-1}$. Let us search for other power-law solutions $n(\omega_k) = k^{-s} = \omega^{-s/\alpha}$. Since integrals over a power-law function generally diverge either at zero or infinite frequency we first check for which s collision integral converges. Physically, convergence means locality of interactions in the frequency space as it signifies that $n(\omega)$ changes only due the waves with the frequencies of order ω.

We have in (1.31) a sum of integrals of power functions, of which every one separately diverges either at $\omega_1 \to 0$ or at $\omega_1 \to \infty$. Cancelations of leading divergencies (by one power at infinity and by two powers at zero) may provide for convergence. Assume $|V(k, k_1, k_2)|^2 \propto k_1^{m_1} k^{2m-m_1}$ at $k_1 \ll k$. Consider $\omega_1 \to \infty$ when $n(\omega_1 - \omega) - n(\omega_1) \approx \omega \partial n_1/\partial \omega_1 \propto n_1 \omega/\omega_1$, $|V(\omega_1, \omega, \omega_1 - \omega)|^2 \propto \omega_1^{(2m-m_1)/\alpha}$ and

$Q(\omega_1, \omega, \omega_1 - \omega) \propto \omega_1^{-2+(2m-m_1+d+1)/\alpha}$. The most dangerous (second) term in (1.31) converges when $s > s_2 = 2m - m_1 + d + 1 - 2\alpha$.

At small frequencies, one should take into account not only the contribution of small ω_1 and $\omega - \omega_1$ in the first term but also of small $\omega_1 - \omega$ in the second term of (1.31):

$$\left(\int_0^\epsilon + \int_{\omega-\epsilon}^\omega \right) Q(\omega, \omega_1, \omega - \omega_1) \{ n_1 n(\omega - \omega_1) - n(\omega)[n_1 + n(\omega - \omega_1)] \} d\omega_1$$

$$-2 \int_\omega^{\omega+\epsilon} d\omega_1 Q(\omega_1, \omega, \omega_1 - \omega)[n(\omega)n(\omega_1 - \omega) - n(\omega_1)(n(\omega) + n(\omega_1 - \omega))]$$

$$= 2 \int_0^\epsilon \omega^\gamma f\left(\frac{\omega}{\omega_1}\right) [n(\omega_1)n(\omega - \omega_1) - n(\omega)(n(\omega_1) + n(\omega - \omega_1))] d\omega_1$$

$$-2 \int_0^\epsilon (\omega_1 + \omega)^\gamma f\left(\frac{\omega_1 + \omega}{\omega_1}\right) [n(\omega)n(\omega_1) - n(\omega_1 + \omega)(n(\omega) + n(\omega_1))] d\omega_1$$

$$= 2 \int_0^\epsilon \left[n(\omega_1)\Big(n(\omega - \omega_1) + n(\omega + \omega_1) - 2n(\omega) \Big) + n(\omega)\Big(n(\omega + \omega_1) \right.$$

$$\left. -n(\omega - \omega_1) \Big) \right] \omega^\gamma f\left(\frac{\omega}{\omega_1}\right) d\omega_1 + 2 \int_0^\epsilon \left[(\omega_1 + \omega)^\gamma f\left(\frac{\omega_1 + \omega}{\omega_1}\right) - \omega^\gamma f\left(\frac{\omega}{\omega_1}\right) \right.$$

$$\left. \times \left[n(\omega_1 + \omega)\left(n(\omega) + n(\omega_1)\right) - n(\omega)n(\omega_1) \right] d\omega_1. \right. \tag{1.32}$$

The integrals converge if $s < s_1 = m_1 + d - 1 + 2\alpha$. Thus, if

$$s_1 > s_2, \quad 2m_1 > 2m + 2 - 4\alpha, \tag{1.33}$$

then there exists an interval of exponents $2m - m_1 + d + 1 - 2\alpha < s < m_1 + d - 1 + 2\alpha$ such that on $n(\omega) \propto \omega^{-s/\alpha}$ the collision integral converges. The cancelations (one at infinity and two at zero) that provide for the "locality interval" are property of the kinetic equation, they generally do not take place for higher nonlinear corrections, at least, nobody was able so far to make re-summations of the perturbation series to have such a locality order-by-order. The exponent of Zakharov distribution $s_0 = m + d$ lies exactly in the middle of the "locality interval": $s_0 = (s_1 + s_2)/2$ when it exists. That means that for the constant-flux distribution the contributions to interactions of all scales, from small to large ones, are counterbalanced and the collision integral in fact vanishes. To show this we use the transformation invented independently by Zakharov and Kraichnan. Let us substitute $n(\omega) = A\omega^{-s/\alpha}$ into (1.31) and make the change of variables $\omega_1 = \omega\omega/\omega_1'$, $\omega_2 = \omega_2'\omega/\omega_1'$ in the second term and do the same, interchanging $1 \leftrightarrow 2$, in the third term:

$$I(\omega) = \int\int_0^\infty d\omega_1 d\omega_2 \left[1 - (\omega/\omega_1)^{\vartheta-1} - (\omega/\omega_2)^{\vartheta-1}\right] R(\omega, \omega_1, \omega_2)$$

$$= A^2\omega^\gamma \int_0^\omega \left[1 - \left(\frac{\omega}{\omega_1}\right)^{\vartheta-1} - \left(\frac{\omega}{\omega-\omega_1}\right)^{\vartheta-1}\right]\left[1 - \left(\frac{\omega}{\omega_1}\right)^{-s/\alpha} - \left(\frac{\omega}{\omega-\omega_1}\right)^{-s/\alpha}\right]$$

$$\times \left[\omega_1(\omega-\omega_1)\right]^{-s/\alpha} f\left(\frac{\omega}{\omega_1}\right) d\omega_1, \tag{1.34}$$

where $\vartheta = 2(m + d - s)/\alpha$. The transformation interchange 0 and ∞ so they are legitimate only for converging integrals. Since $f(x)$ is a positive function, the integrals become zero only at $s_0 = \alpha$ and $s_1 = m+d$. Therefore, the Rayleigh-Jeans and Kolmogorov-Zakharov distributions are the only universal power-law stationary solutions of the kinetic equation. Each of these solutions is a one-parameter solution in the isotropic case. Rayleigh-Jeans equipartition describes equilibrium which takes place for the closed system of waves, that is at zero forcing and dissipation: $I_k^3\{k^{-s_0}\} = 0$. On the contrary, Kolmogorov-Zakharov solution has singularity at $\omega = 0$ (which corresponds to a source): $I_k^3\{k^{-s_1}\} \propto \delta(\omega)$. Let us show that indeed Kolmogorov-Zakharov solution provides a constant flux of energy in k-space: $P_k = -\int_{k'<k} d\mathbf{k}'\omega_{k'} I_{k'}^{(3)} = -\int_0^k \pi(2k')^{d-1}\omega_{k'} I_{k'}^{(3)} dk'$. Using (1.31) we find that $P(\omega) \equiv P_{k(\omega)}$ is given by $P(\omega) = -\int_0^\omega \omega' I(\omega')d\omega'$. For the exponents s from to the locality interval one can consider the collision integral as a function of s. Passing in (1.34) to $y = \omega_1/\omega$ one finds $I(\omega) \propto \omega^{\vartheta-2} A^2 I(s)$ and

$$I(s) = \int_0^1 dy \left[1 - y^{1-\vartheta} - (1-y)^{1-\vartheta}\right]\left[1 - y^{s/\alpha} - (1-y)^{s/\alpha}\right][y(1-y)]^{-s/\alpha} f(1/y).$$

Hence, the flux on power-law distributions is as follows: $P = -\omega^\vartheta A^2 I(s)/\vartheta$. At $s = s_0 = m + d$ there is an indeterminacy of the form $0/0$ since $I(m+d) = 0$ and $\vartheta(s_0) = 0$. Using the L'Hospital's rule, we obtain an expression where the energy flux is proportional to the derivative of the collision integral with respect to the exponent: $P = -2A^2 I'(s_0)/\alpha$

$$P = -A^2 \int_0^1 [y\ln y + (1-y)\ln(1-y)]\left[y^{s_0/\alpha} + (1-y)^{s_0/\alpha} - 1\right]\frac{f\left(y^{-1}\right)dy}{[y(1-y)]^{s_0/\alpha}}.$$

The sign of P coincides with that of $1 - y^{s_0/\alpha} - (1 - y)^{s_0/\alpha}$, which is that of $s_0/\alpha - 1$, i.e. the equilibrium exponent $s_0 = \alpha$ provides the boundary between the positive and the negative fluxes to large k. Indeed, flux must flow trying to restore equipartition. We conclude that the Kolmogorov-Zakharov spectrum describes physical turbulent state provided it decays faster than the equilibrium. In the opposite case, $s_0 < \alpha$, a flux solution is not physical, it is actually unstable (Zakharov et al 1992).

(ii) **Exercise 1.2** Consider an anisotropic forcing that produces non-vanishing rate of injection of momentum into the system: $\int \mathbf{k} F_k d\mathbf{k} \equiv \mathbf{R}$. Using dimensional analysis we can write for a steady state (assuming it exists)

$$n(\mathbf{k}, P, \mathbf{R}) = \lambda P^{1/2} k^{-m-d} f(\xi), \quad \xi = \frac{(\mathbf{R} \cdot \mathbf{k}) \omega(k)}{P k^2} , \qquad (1.35)$$

where λ is the dimensional Kolmogorov constant; the medium is assumed to be isotropic, therefore the solution depends on $(\mathbf{R} \cdot \mathbf{k})$. The dimensionless function $f(\xi)$ has been found analytically only for sound with positive dispersion (Zakharov et al 2001). In a general case, one can *assume* that $f(\xi)$ is analytical at zero; then, expanding (1.35), one obtains a stationary anisotropic correction to the isotropic solution

$$\begin{aligned} n(\mathbf{k}, \mathbf{P}, \mathbf{R}) &\approx \lambda P^{1/2} k^{-m-d} + \lambda f'(0) k^{-m-d} (\mathbf{R} \cdot \mathbf{k}) \omega(k) P^{-1/2} k^{-2} \\ &= n_0(k) + \delta n(\mathbf{k}) . \end{aligned} \qquad (1.36)$$

The ratio $\delta n / n_0 \propto \omega(k)/k$ increases with k for waves with the decay dispersion law (when the three-wave kinetic equation is relevant). That is the spectrum of the weak turbulence generated by weakly anisotropic pumping is getting more anisotropic as we go into the inertial interval of scales.

To verify that (1.36) is the stationary solution of the linearized kinetic equation, let us substitute $n(\mathbf{k}) = k^a [1 + k^b (\mathbf{q} \mathbf{R})]$ with $\mathbf{q} = \mathbf{k}/k$ into the linearized collision integral, which may be represented as

$$\hat{L}_k \delta n(\mathbf{k}) = \mathbf{R} \cdot \mathbf{I}'(\mathbf{k}),$$

$$\mathbf{I}'(\mathbf{k}) = \int d\mathbf{k}_1 d\mathbf{k}_2 [U(\mathbf{k}, \mathbf{k}_1, \mathbf{k}_2)\mathbf{f}(\mathbf{k}, \mathbf{k}_1, \mathbf{k}_2) - U(\mathbf{k}_1, \mathbf{k}_2, \mathbf{k})\mathbf{f}(\mathbf{k}_1, \mathbf{k}_2, \mathbf{k})$$

$$-U(\mathbf{k}_2, \mathbf{k}, \mathbf{k}_1)\mathbf{f}(\mathbf{k}_2, \mathbf{k}, \mathbf{k}_1)], \tag{1.37}$$

$$U(\mathbf{k}, \mathbf{k}_1, \mathbf{k}_2) = U_{k12} = \pi |V(k, k_1, k_2)|^2 \delta(\omega_k - \omega_1 - \omega_2)\delta(\mathbf{k} - \mathbf{k}_1 - \mathbf{k}_2),$$

$$\mathbf{f}(\mathbf{k}, \mathbf{k}_1, \mathbf{k}_2) = -\mathbf{q}k^{a+b}(k_1^a + k_2^a) + \mathbf{q}_1 k_1^{a+b}(k_2^a - k^a) + \mathbf{q}_2 k_2^{a+b}(k_1^a - k^a) \ .$$

Isotropy allows us to write $\mathbf{I}' = \mathbf{q}I(\mathbf{k})$ where

$$I = \int d\mathbf{k}_1 d\mathbf{k}_2 [U_{k12}\mathbf{f}(\mathbf{k}, \mathbf{k}_1, \mathbf{k}_2) - U_{12k}\mathbf{f}(\mathbf{k}_1, \mathbf{k}_2, \mathbf{k}) - U_{2k1}\mathbf{f}(\mathbf{k}_2, \mathbf{k}, \mathbf{k}_1)] \ .$$

Similarly to what we did in the Problem 1, let us look for the change of the integration variables that transform the second and the third terms of the integral into the first one (with certain factors). Such transformation was found out by Kats and Kontorovich (see Zakharov et al 1992 and the references therein). Let us relabel the integration variables so that the third term takes the form $\int d\mathbf{k}_1' d\mathbf{k}_2' U(\mathbf{k}_2', \mathbf{k}, \mathbf{k}_1') \mathbf{q} \cdot \mathbf{f}(\mathbf{k}_2', \mathbf{k}, \mathbf{k}_1')$. The integrand is non-vanishing only for those $\mathbf{k}_i'$ which satisfy the conservation of energy and momentum. Figure 1.9, a and c show triangles corresponding to the momentum conservation in the first and third term respectively.

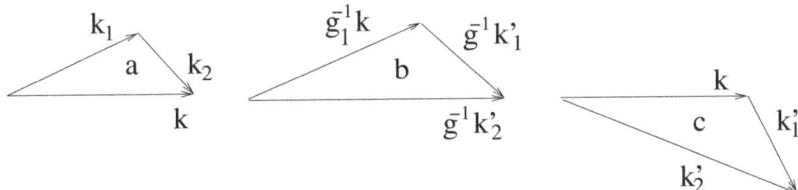

Fig. 1.9. Transformation which converts two triangles (c and a) into one another. The triangles express the laws of conservation of energy and momentum.

The two triangles have a common vector $\mathbf{k}$ and the map c→a is the combination of rotation $\hat{g}_1^{-1}$, depicted in Figure c in 1.9 which maps the original triangle to the re-scaled desired triangle and then re-scaling $\hat{\lambda}$ with the coefficient $\lambda_1 = k/k_1 = k_2'/k$ that finishes the transformation. Such map means the change of variables $\mathbf{k}_2' = (\lambda_1 \hat{g}_1)^2 \mathbf{k}_1$ and $\mathbf{k}_1' = \lambda_1 \hat{g}_1 \mathbf{k}_2$. Taking into account the Jacobian and using

$\mathbf{k} = \lambda_1 \hat{g}_1 \mathbf{k}_1$ we find

$$\int d\mathbf{k}_1' d\mathbf{k}_2' U(\mathbf{k}_2', \mathbf{k}, \mathbf{k}_1') \mathbf{q} \cdot \mathbf{f}(\mathbf{k}_2', \mathbf{k}, \mathbf{k}_1') = \int d\mathbf{k}_1 d\mathbf{k}_2 U(\lambda_1 \hat{g}_1 \mathbf{k}, \lambda_1 \hat{g}_1 \mathbf{k}_1, \lambda_1 \hat{g}_1 \mathbf{k}_2}$$
$$\times \mathbf{q} \cdot \mathbf{f}(\lambda_1 \hat{g}_1 \mathbf{k}, \lambda_1 \hat{g}_1 \mathbf{k}_1, \lambda_1 \hat{g}_1 \mathbf{k}_2) \lambda_1^{3d} \,.$$

Since $U(\lambda_1 \hat{g}_1 \mathbf{k}, \lambda_1 \hat{g}_1 \mathbf{k}_1, \lambda_1 \hat{g}_1 \mathbf{k}_2) = \lambda_1^{2m-d-\alpha} U(\mathbf{k}, \mathbf{k}_1, \mathbf{k}_2)$ and

$$\mathbf{q} \cdot \mathbf{f}(\lambda_1 \hat{g}_1 \mathbf{k}, \lambda_1 \hat{g}_1 \mathbf{k}_1, \lambda_1 \hat{g}_1 \mathbf{k}_2) = \lambda_1^{2a+b} \mathbf{q} \cdot \hat{g}_1 \mathbf{f}(\mathbf{k}, \mathbf{k}_1, \mathbf{k}_2) = \lambda_1^{2a+b} \mathbf{q}_1 \cdot \mathbf{f}(\mathbf{k}, \mathbf{k}_1, \mathbf{k}_2),$$

$\int d\mathbf{k}_1' d\mathbf{k}_2' U(\mathbf{k}_2', \mathbf{k}, \mathbf{k}_1') \mathbf{q} \cdot \mathbf{f}(\mathbf{k}_2', \mathbf{k}, \mathbf{k}_1') = \lambda^w \int d\mathbf{k}_1 d\mathbf{k}_2 U(\mathbf{k}, \mathbf{k}_1, \mathbf{k}_2) \mathbf{q}_1 \cdot \mathbf{f}(\mathbf{k}, \mathbf{k}_1, \mathbf{k}_2$
with $w = 2m + 2d - \alpha + 2a + b$. Interchanging 1 and 2 we obtain $I(\mathbf{k})$
in the factorized form:

$$I(\mathbf{k}) = \int d\mathbf{k}_1 d\mathbf{k}_2 U(\mathbf{k}, \mathbf{k}_1, \mathbf{k}_2) \mathbf{f}(\mathbf{k}, \mathbf{k}_1, \mathbf{k}_2) \left[\mathbf{q} - \mathbf{q}_1 (k/k_1)^w - \mathbf{q}_2 (k/k_2)^w \right]$$

When we choose $w = 2m + 2d - \alpha + 2a + b = -1$, then $I(\mathbf{k}) = 0$
due to the δ-function of wave vectors. It is easily seen that the index
obtained corresponds to δn from (1.36). To conclude that the drift
solutions are steady modes, we have to verify their locality, i.e., the
convergence of $I(\mathbf{k})$ which is done similarly to the isotropic case in the
Problem 1. The only difference is that the divergences are reduced
by the power of k rather than ω_k, due to which the locality strip is
compressed by $2(\alpha - 1)$.

The direction of the momentum flux (to large or small k) is given
by the sign of the collision integral derivative with respect to the
index of the solution: $\operatorname{sign} R = -\operatorname{sign}(\partial I / \partial b)$.

(iii) **Exercise 1.3** Our consideration here is more intuitive and less formal
than that of Gawędzki, Sect. 3.4.2. Since $\dot{W} = \sigma W$ then $\dot{I} = \sigma I + I \sigma^T$. We assume the following ordering of the eigenvalues $\rho_1 \geq \rho_2 \geq \dots \geq \rho_d$. The equation on I becomes

$$\partial_t \rho_i = \tilde{\sigma}_{ii}, \quad \tilde{\sigma} = O \sigma O^T, \quad \partial_t O = \Omega O, \quad \Omega_{ij} = \frac{e^{2\rho_i} \tilde{\sigma}_{ji} + e^{2\rho_j} \tilde{\sigma}_{ij}}{e^{2\rho_i} - e^{2\rho_j}},$$
$$\tag{1.38}$$

where $\rho_i(0) = 0$ and $O_{ij}(0) = \delta_{ij}$. Here and below we do not sum
over repeated indices unless stated otherwise. Note that Ω is antisym-
metric to preserve $O^T O = 1$. We assume the spectrum of Lyapunov
exponents to be non-degenerate, then at times much larger than the

maximum of $(\lambda_i - \lambda_{i+1})^{-1}$ we have $\rho_1 \gg \rho_2 \gg \ldots \gg \rho_d$ and matrix Ω becomes independent of ρ_i

$$\Omega_{ik} = \tilde{\sigma}_{ki}, \quad i < k, \quad \Omega_{ik} = -\tilde{\sigma}_{ik}, \quad i > k. \tag{1.39}$$

The above independence allows us to resolve explicitly the equation on ρ_i as follows

$$\rho_i(t) = \int_0^t \tilde{\sigma}_{ii}(t')dt', \quad \lambda_i = \lim_{t \to \infty} \frac{1}{t} \int_0^t \tilde{\sigma}_{ii}(t')dt'. \tag{1.40}$$

The above representation of Lyapunov exponents is a (non-rigorous) proof that the limits defining Lyapunov exponents exist as Oseledec theorem states. Indeed let us make the (ergodic) hypothesis that the above time-average can be calculated as an average over the statistics of velocity field. Then, provided the statistics of $\tilde{\sigma}_{ii}(t)$ becomes stationary at large t, the Lyapunov exponents λ_i become the subject of the law of large numbers and we have

$$\lambda_i = \lim_{t \to \infty} \langle \tilde{\sigma}_{ii}(t) \rangle = \lim_{t \to \infty} \sum_{jk} \langle O_{ij}(t)O_{ik}(t)\sigma_{jk}(t) \rangle. \tag{1.41}$$

While generally the analytic calculation of that average is not possible, it is readily accomplished in a statistically isotropic case when the correlation time τ of σ is small comparing to the rms value σ_c. For Δt satisfying $\tau \ll \Delta t \ll \sigma_c^{-1}$ we have $O_{ij}(t) = O_{ij}(t - \Delta t) + \int_{t-\Delta t}^t \Omega_{ik}(t')O_{kj}(t')dt' + o(\sigma_c \Delta t)$ where $O_{ij}(t - \Delta t)$, being determined by $\sigma(t)$ at times smaller than $t - \Delta t$, is approximately independent of $\sigma(t)$ by $\Delta t \gg \tau$. First iteration of O_{ij} gives at large time

$$\lambda_i = \langle O_{ij}(t - \Delta t)O_{ik}(t - \Delta t)\sigma_{jk}(t) \rangle + \sum_{jk} \int_{t-\Delta t}^t dt' \Big\langle \sigma_{jk}(t)$$

$$\times \Big[\Omega_{il}(t')O_{lj}(t')O_{ik}(t - \Delta t) + \Omega_{il}(t')O_{lk}(t')O_{ij}(t - \Delta t)\Big] \Big\rangle + o(\sigma_c \Delta t),$$

Introducing $\tilde{\sigma}'(t) = O(t - \Delta t)\sigma(t)O^T(t - \Delta t)$, performing second iteration of $O_{ij}(t')$ and using (1.39) and the symmetry of thr indices we find

$$\lambda_i = \langle \tilde{\sigma}'_{ii} \rangle + \int_{t-\Delta t}^t dt' \Big[\sum_{l>i} \langle (\tilde{\sigma}'_{il}(t) + \tilde{\sigma}'_{li}(t))\tilde{\sigma}'_{il}(t') \rangle - \sum_{l<i} \langle (\tilde{\sigma}'_{il}(t) + \tilde{\sigma}'_{li}(t))\tilde{\sigma}'_{il}(t') \rangle \Big].$$

For $\Delta t \gg \tau$ the matrix $\sigma_{ij}(t)$ is independent of $O_{ij}(t - \Delta t)$ which is determined by σ at times earlier than $t - \Delta t$. Due to isotropy, the matrix $\tilde{\sigma}'$ has the same statistics as σ and can be replaced by it in the correlation functions. The first average in is independent of i and equals $\langle tr\sigma \rangle / d = \sum \lambda_i / d$ while the second is independent of l so that

$$\lambda_i = d^{-1} \sum_p \lambda_p + (d - 2i + 1) \int_{t-\Delta t}^t dt' \langle (\sigma_{il}(t) + \sigma_{li}(t)) \sigma_{il}(t') \rangle ,$$

where there is no summation over the repeated indices in the second term (note that summation over i produces identity). We may write

$$\int_{t-\Delta t}^t dt' \langle \sigma_{ij}(t) \sigma_{kl}(t') \rangle = \int_{t-\Delta t}^t dt' \langle\langle \sigma_{ij}(t) \sigma_{kl}(t') \rangle\rangle + \Delta t \langle \sigma_{ij} \rangle \langle \sigma_{kl} \rangle$$

where double brackets stand for dispersion. Noting from $\langle \sigma_{jk} \rangle = \sum_p \lambda_p \delta_{jk} / d$ that the last term contains additional factor $\sum \lambda_i \Delta t \ll \sigma_c \tau \ll 1$ with respect to the first term we conclude that it can be neglected. Here we used $\sum \lambda_i \sim \sigma_c^2 \tau$ and $\Delta t \ll \sigma_c^{-1}$. Using stationarity of the statistics of $\sigma_{ij}(t)$ at large times we may write

$$\lambda_i = d^{-1} \sum \lambda_p + (d - 2i + 1)(C_{ijij} + C_{ijji})/2,$$

$$= -(2d)^{-1} \sum_{j,l} C_{jjll} + (d - 2i + 1)(C_{ikik} + C_{kiik})/2. \quad (1.42)$$

$$C_{ijkl} = \lim_{t \to \infty} \int dt' \langle\langle \sigma_{ij}(t) \sigma_{kl}(t') \rangle\rangle ,$$

where there is no summation in the last term in λ_i. Note the symmetry $C_{ijkl} = C_{ilkj}$. Due to isotropy and the symmetry $C_{ijkl} + C_{ilkj}$ we have $C_{ijkl} = A\delta_{ik}\delta_{jl} + B(\delta_{ij}\delta_{kl} + \delta_{il}\delta_{jk})$, where we also assumed parity. It is convenient to write the two remaining constants as

$$C_{ijkl} = 2D_1 \left[(d + 1 - 2\Gamma)\delta_{ik}\delta_{jl} + (\Gamma d - 1)(\delta_{ij}\delta_{kl} + \delta_{il}\delta_{jk}) \right] .$$
$$(1.43)$$

Then it is easy to show using dispersion non-negativity conditions that D_1 and Γ are non-negative. While D_1 measures the overall rate of strain, Γ is the degree of compressibility, it changes between zero (for an incompressible flow) and unity (for a potential flow). Since Γ vanishes for incompressible flow one has $\sum \lambda_i \propto \Gamma$:

$$\sum \lambda_i = -\sum_{jl} C_{jjll}/2 = -\Gamma D_1 d(d-1)(d+2). \quad (1.44)$$

The final answer takes the form

$$\lambda_i = D_1 \left[d(d - 2i + 1) - 2\Gamma(d + (d-2)i) \right] . \quad (1.45)$$

Formulas (1.43) and (1.45) correspond respectively to (2.9) and (2.53) from Gawędzki. The senior Lyapunov exponent, $\lambda_1 = D_1(d-1)[d - 4\Gamma]$ decreases linearly when compressibility degree grows. Thus the effect of compressibility is to suppress the exponential divergence of nearby trajectories. For an incompressible random flow where $\Gamma = 0$, the first Lyapunov exponent is positive. Generally $\lambda_1 \geq 0$ for incompressible flow because volume conservation implies $\sum \lambda_i = 0 \leq \lambda_1$. On the hand in the case $d = 1$ where compressibility is always maximal $\Gamma = 1$ (the only incompressible flow in one dimension is a constant one) we always have $\lambda_1 < 0$ (to define this limit one should assume that $D_1(d - 1)$ is a finite constant). In dimensions 2 and 3, Γ becomes negative at the critical compressibility $\Gamma_{cr} = d/4$. Finally λ_1 is always positive at $d > 4$ while in four dimensions $\lambda_1 > 0$ unless the flow is potential where $\lambda_1 = 0$.

References

Antonsen, T. and Ott, E. (1991). Multifractal power spectra of passive scalars convected by chaotic fluid flows, *Phys. Rev. A* **44**, 851–871.

Balkovsky, E. and Lebedev, V. (1998). Instanton for the Kraichnan Passive Scalar Problem, *Phys. Rev. E* **58**, 5776–5786.

Balkovsky, E. and Fouxon, A. (1999). Universal long-time properties of Lagrangian statistics in the Batchelor regime and their application to the passive scalar problem. *Phys. Rev. E* **60**, 4164–4174.

Batchelor, G. K. (1959). Small-scale variation of convected quantities like temperature in turbulent flows, *J. Fluid Mech.* **5**, 113–120.

Bauer, M. and Bernard, D. (2006). 2d growth processes: SLE and Loewner chains, *Phys. Reports* **432**, 115.

Bernard, D., Gawędzki, K. and Kupiainen, A. (1996). Anomalous scaling of the N-point functions of a passive scalar, *Phys. Rev. E* **54**, 2564–2567.

Bernard, D., Boffetta, G., Celani, A. and. Falkovich, G. (2006). Conformal invariance in two-dimensional turbulence, Nature Physics **2**, 124–128.

Boffetta, G. Celani, A. and Vergassola, M. (2000). Inverse energy cascade in two-dimensional turbulence: Deviations from Gaussian behavior, *Phys. Rev. E* **61** R29–R32.

Cardy, J. (2005). SLE for theoretical physicists. *Ann. Phys.* **318**, 81–118.

Celani, A., Lanotte, A. Mazzino, A. and Vergassola, M. (2001). Fronts in passive scalar turbulence, *Phys. Fluids* **13**, 1768–1783.

Chen, S. et al, (2006). Physical Mechanism of the Two-Dimensional Inverse Energy Cascade, *Phys. Rev. Lett.* **96**, 084502.

Chertkov, M., Falkovich, G., Kolokolov, I. and Lebedev, V. (1995). Normal and anomalous scaling of the fourth-order correlation function of a randomly advected scalar", *Phys. Rev. E* **52**, 4924–4934.

Chertkov, M. and Falkovich, G. (1996). Anomalous scaling exponents of a white-advected passive scalar, *Phys. Rev Lett.* **76**, 2706–2709.

Chertkov, M. Kolokolov, I. and Vergassola, M. (1998). Inverse versus direct cascades in turbulent advection, *Phys. Rev. Lett.* **80**, 512-515.

Corrsin, S. (1952). Heat transfer in isotropic turbulence, *J. Appl. Phys.* **23**, 113-116.

Gawędzki, K. and Vergassola, M. (2000). Phase transition in the passive scaler advection, *Physica D* **138**, 63–90.

Choi, Y. Lvov, Y. and Nazarenko, S. (2005). Joint statistics of amplitudes and phases in wave turbulence,*Physica D* **201**, 121–149.

Dyachenko, A. and Falkovich, G. (1996). Condensate turbulence in two dimensions, *Phys. Rev. E* **54**, 5095–5098.

Dyachenko, A. Newell, A. Pushkarev A. and Zakharov, V. (1992). Optical turbulence, Physica D **57**, 96–120.

E, W. , Khanin, K., Mazel, A. and .Sinai, Ya.G. (1997). *Phys. Rev. Lett.* **78**, 1904-1907.

Falkovich, G. and Lebedev, V. (1994). Universal direct cascade in two-dimensional turbulence, *Phys. Rev. E* **50**, 3883–3899.

Falkovich, G., Gawędzki, K. and Vergassola, M. (2001). Particles and fields in fluid turbulence, *Rev. Mod. Phys.*, **73**, 913–975.

Falkovich G. and Sreenivasan, K. (2006). Lessons from hydrodynamic turbulence, *Physics Today*, **59**, 43–49.

Frisch, U. (1995). *Turbulence* (Cambridge Univ. Press, Cambridge,).

Frisch U. and Bec, J. (2001). Burgulence, in *Les Houches 2000: New Trends in Turbulence*, ed. M. Lesieur, (Springer EDP-Sciences, Berlin)

Gawędzki, K. and Kupiainen, A. (1995). Anomalous scaling of the passive scalar, *Phys. Rev. Lett.* **75**, 3834-3837.

Gruzberg, I. and Kadanoff, L. (2004). The Loewner equation: maps and shapes. *J. Stat. Phys.* **114**, 1183–1198.

Kellay, H. and Goldburg, W., (2002). Two-dimensional turbulence: a review of some recent experiments. *Rep. Prog. Phys.* **65**, 845–894.

Kolmogorov, A. N. (1941). The local structure of turbulence in incompressible viscous fluid for very large Reynolds number, *C. R. Acad. Sci. URSS* **30**, 301–305.

Kraichnan, R. H. (1967). Inertial ranges in two-dimensional turbulence, *Phys. Fluids* **10**, 1417–1422.

Kraichnan, R. H. (1968). Small-scale structure of a scalar field convected by turbulence, *Phys. Fluids* **11**, 945–963.

Kraichnan, R. H. (1974). Convection of a passive scalar by a quasi-uniform random straining field, *J. Fluid Mech.* **64**, 737–750.

Kuznetsov, E. A. (2004). Turbulence Spectra Generated by Singularities, *JETP Letters* **80**, 83–86.

Landau, L. and Lifshits, E. (1987). *Fluid Mechanics* (Pergamon Press, Oxford)

Lawler, G. (2005). Conformally invariant processes in the plane, *Mathematical Surveys and Monographs* **114**, 1183–1198.

Obukhov, A. M. (1949). Structure of the temperature field in turbulent flows, *Izv. Akad. Nauk SSSR, Geogr. Geofiz.* **13**, 58–61.

Phillips, O. (1977). *The Dynamics of the Upper Ocean* (Cambridge Univ. Press, Cambridge).

Polyakov, A. M. (1970). Conformal symmetry of critical fluctuations, *JETP Lett.* **12**, 381–383.

Schramm, O. (2000). Scaling limits of loop-erased random walks and uniform spanning trees. *Israel J. Math.* **118**, 221–288.

Shraiman, B. and Siggia, E. (1995). Anomalous scaling of a passive scalar in turbulent flow", *C.R. Acad. Sci.*, **321**, 279–285

Tabeling, P. (2002). Two-dimensional turbulence: a physicist approach, *Phys. Rep.* **362**, 1–62.

Zakharov, V., Lvov, V. and Falkovich, G. (1992). *Kolmogorov spectra of turbulence* (Springer, Berlin)

2

Krzysztof Gawędzki. Soluble models of turbulent transport

2.1 Introduction

Transport of matter, pollution or chemical and biological agents by turbulent flows is an important phenomenon with multiple applications from cosmology and astrophysics to meteorology, environmental studies, biology, chemistry and engineering. This is a series of lecture notes reviewing some aspects of the theoretical work on simple models of turbulent transport done over the last decade and partly described already, often with much more details, in [Fal01]. Other approaches to modeling turbulent transport, with the stress on turbulent diffusion, may be found in the review [Maj99]. More practical issues related to the influence of turbulence on the chemical or biological activity were addressed in [Tel05] from somewhat similar point of view.

The aim of simple models of turbulent transport that we shall discuss here is to explain or discover general phenomena and robust behaviors rather than to provide a detailed quantitative description. We shall study exclusively the passive transport approximation which is relevant for small concentrations of transported matter so that its back-reaction on the flow itself may be ignored. From the mathematical point of view, passive transport may be considered as a problem in random dynamical systems. Some of the techniques developed over years in the dynamical systems theory are indeed useful in analyzing transport phenomena, especially for flows with moderate Reynolds numbers. Conversely, an interest in transport phenomena calls for specific developments in the dynamical systems theory. In particular, dealing with high Reynolds number flows requires new tools that pertain to the regime where velocities are rough. Mathematical study of the corresponding dynamical systems is a rather new field.

These notes are organized as follows. In Lecture 1, we describe two ways

44

in which turbulent flow may be viewed as a dynamical system, with the second way relevant for the turbulent transport problem. We subsequently discuss relations between passive transport of local objects and advection of fields. Lecture 2 is devoted to the aspects of dynamical systems that are important in studying transport phenomena. In particular, we discuss the concept of natural invariant measure and the multiplicative ergodic theory, with the stress on multiplicative large deviations. Multiplicative versions of fluctuation relations of the Evans-Searles and the Gallavotti-Cohen types, pertaining to compressible flows, are also described. Lecture 3 introduces the Kraichnan model of turbulence. We describe how the statistics of multiplicative large deviations may be obtained in this model and give an example of a transport problem which may be solved exactly employing this result. In Lecture 4, we analyze Kraichnan flows that model high Reynolds number turbulence. We discuss different exotic behaviors of flow trajectories that occur in such flows, depending on the degree of compressibility. We indicate how such behaviors lead to different cascade-like non-equilibrium states for advected fields. Finally, in Lecture 5, we describe the so called "zero mode" mechanism behind intermittency and anomalous scaling in turbulent transport. Discovered in the Kraichnan model, this mechanism is related to hidden statistical conservation laws, see Sect. 1.5 of the course of G. Falkovich. In End Remarks, we briefly summarize what has been learned from simple models about turbulent transport and point to some open questions. Each lecture is followed by a problem set designed to facilitate assimilation of the presented material. Solutions of the problems may be found in Sect. 2.9.

As far as the prerequisites for these lectures are concerned, we assume some rudimentary knowledge of basic facts about ordinary differential equations and of basic notions from probability theory, as well as some familiarity with the Brownian motion and the white noise processes. Mathematical arguments used in the text are often based on manipulating stochastic differential equations. The reader not familiar with that concept should study Problem 1.1 after Lecture 1, and its solution, before reading Lectures 3-5.

2.2 Lecture 1. Turbulent flow as a dynamical system

There are two different ways in which a turbulent flow may be viewed as a **dynamical system** [Boh98]. The first one is related to hydrodynamical equations, like the Navier-Stokes ones, that govern the evolution of the velocity field and, eventually, of other relevant fields like pressure, fluid density, temperature, etc. The second way, that will be the main subject of these lectures, is related to transport phenomena.

2.2.1 Navier-Stokes equations

These equations, dating back to the work of Claude-Louis Navier from 1823 and of George Gabriel Stokes from 1843, have the form of the Newton equation for the fluid element. In the simplest case of incompressible fluid, they read:

$$\rho\left(\partial_t \boldsymbol{v} + (\boldsymbol{v} \cdot \boldsymbol{\nabla})\boldsymbol{v} - \nu \boldsymbol{\nabla}^2 \boldsymbol{v}\right) = -\boldsymbol{\nabla} p + \boldsymbol{f}. \qquad (2.1)$$

fluid density   kinematical viscosity   pressure   force density

The above equation has to be supplemented by the incompressibility conditions $\rho = \text{const.}$ and $\boldsymbol{\nabla} \cdot \boldsymbol{v} = 0$. Altogether, they define an evolution equation on the (infinite dimensional) space $\mathscr{V}_0$ of the divergenceless vector fields. This equation may be viewed as an infinite-dimensional (non-autonomous) dynamical system of the form

$$\frac{\mathrm{d}X}{\mathrm{d}t} = \mathscr{X}(t, X) \qquad \text{for } X \in \mathscr{V}_0.$$

If $\boldsymbol{f}$ is random then the right hand side $\mathscr{X}(t, X)$ will be random describing an infinite dimensional random dynamical system. Clearly, eq. (2.1) is non-linear. The strength of the non-linear term $(\boldsymbol{v} \cdot \boldsymbol{\nabla})\boldsymbol{v}$, relative to the linear one $\nu \boldsymbol{\nabla}^2 \boldsymbol{v}$ describing the viscous dissipation, depends on the length scale l. It is captured by the running Reynolds number defined as

$$Re(l) = \frac{\Delta_l v \cdot l}{\nu},$$

where $\Delta_l v$ is the size of a typical velocity difference across distance l. In particular, for L being the **integral scale** corresponding to the size of the flow recipient, $Re(L) \equiv Re$ is the integral scale Reynolds number. On the other end, $Re(\eta) = 1$ for η equal to the **viscous** (or Kolmogorov) **scale**. At scale η, the non-linear and the dissipative terms in the equation are of comparable strengths. Phenomenological observations point to the following classification of flows according to the size of the Reynolds number Re:

(a) $Re \lesssim 1$:   laminar flows,
(b) $Re \sim 10$ to 10^2:   onset of turbulence,
(c) $Re \gtrsim 10^3$:   developed turbulence.

In regime (a), some explicit solutions are known, see e.g. [Bat67]. In regime (b), the flow is driven by few unstable modes. The standard dynamical systems theory studying temporal evolution of few degrees of freedom governed by ordinary differential equations or iterated maps proved useful here, for example in describing the scenarios of appearance of the chaotic motions, see e.g. [Eck81]. Finally, in regime (c), there are many unstable degrees of freedom. Kolmogorov's scaling theory [Kol41] predicts that in this regime,

$$\Delta_l v \propto l^{1/3} \qquad \text{for} \qquad \eta \ll l \ll L,$$

which is not very far from the observed behavior, as discussed in Sect. 1.4.1 of the course of G. Falkovich. The number of unstable modes may be estimated to be of the order of $(L/\eta)^3 \approx Re^{9/4}$. Although the flow may still be analyzed with the use of concepts derived from the standard dynamical systems theory [Bof02], new phenomena arise here, like cascades with (approximately) constant energy flux, intermittency, anomalous corrections to scaling etc., that need a new theory, see the lectures of G. Falkovich in this volume.

2.2.2 Transport phenomena

The second way to view turbulent flow as a dynamical system is related to transport of particles and fields by the fluid.

2.2.2.1 Transport of particles

The particles may be idealized fluid elements, called **Lagrangian particles**, or small objects without inertia suspended in the fluid that also undergo molecular diffusion, see Sect. 3.2 of the course by J. Cardy. We talk in the latter case of

1. **tracer particles**

    whose motion is governed by the evolution equation

    $$\frac{\mathrm{d}\boldsymbol{R}}{\mathrm{d}t} = \boldsymbol{v}(t, \boldsymbol{R}) + \sqrt{2\kappa}\,\boldsymbol{\eta}(t) \tag{2.2}$$

    where κ stands for the diffusivity and $\boldsymbol{\eta}(t)$ is the standard vector-valued white noise. For $\kappa = 0$, one obtains the equation for the Lagrangian particles.

Objects like water droplets in the air or heavy dust particles experience a friction force proportional to their relative velocity with respect to the fluid. These are

2. particles with inertia

whose position $\boldsymbol{R}$ and velocity $\boldsymbol{V}$ satisfy the the equations of motion [Bec05]

$$\frac{\mathrm{d}\boldsymbol{R}}{\mathrm{d}t} = \boldsymbol{V}\,, \quad \frac{\mathrm{d}\boldsymbol{V}}{\mathrm{d}t} = \frac{1}{\tau}\big(-\boldsymbol{V} + \boldsymbol{v}(t,\boldsymbol{R}) + \sqrt{2\kappa}\,\boldsymbol{\eta}(t)\big) \tag{2.3}$$

where τ is the Stokes time. In the $\tau \to 0$ limit the equation for the tracer particles is recovered.

Finally, one may consider bubbles with deformable shape suspended in the fluid, or extended objects with more structure like

3. polymer molecules

whose motion may be modeled by the differential equations [Bir87]

$$\frac{\mathrm{d}\boldsymbol{R}}{\mathrm{d}t} = \boldsymbol{v}(t,\boldsymbol{R}) + \sqrt{2\kappa}\,\boldsymbol{\eta}(t)\,, \tag{2.4}$$

$$\frac{\mathrm{d}\boldsymbol{B}}{\mathrm{d}t} = (\boldsymbol{B}\!\cdot\!\boldsymbol{\nabla})\boldsymbol{v}(t,\boldsymbol{R}) - \alpha\boldsymbol{B} + \sqrt{2\sigma}\,\boldsymbol{\xi}(t)\,, \tag{2.5}$$

where $\boldsymbol{R}$ is the position of one end of the polymer chain and $\boldsymbol{B}$ is the end-to-end separation vector. Here, $\boldsymbol{\eta}(t)$ and $\boldsymbol{\xi}(t)$ are independent white noises. The term $-\alpha\boldsymbol{B}$ describes the elastic force that counteracts the stretching of polymers whose two ends move with different velocities.

All three cases provide (non-autonomous) dynamical systems in space or phase space if $\boldsymbol{v}, \boldsymbol{\eta}, \boldsymbol{\xi}$ are given. The dynamical systems are random if $\boldsymbol{v}, \boldsymbol{\eta}, \boldsymbol{\xi}$ are random with given statistics. Such dynamical systems may be studied with tools of the theory of finite-dimensional dynamical systems. Their specificity has provided, however, a number of new inputs to that theory.

2.2.2.2 Transport of fields

Turbulent flows may also transport physical quantities described by fields defined at each space (or phase-space) point, like temperature, pollutant or dye density, magnetic field, etc. For small field intensity, the time evolution of such transported fields is described by linear advection-diffusion equations into which the fluid velocity enters as variable coefficients. Besides, under the same assumption, one may often ignore the back-reaction of the transported field on the motion of the fluid, ending up in a **passive transport** approximation decoupling the fluid evolution from that of the advected fields. The passive advection of a

1a. scalar field $\theta(t, \boldsymbol{r})$

(e.g. temperature) is governed by the advection-diffusion equation

$$\partial_t \theta + (\boldsymbol{v} \cdot \boldsymbol{\nabla})\theta - \kappa \boldsymbol{\nabla}^2 \theta = g \,, \tag{2.6}$$

where κ is the diffusivity constant of the field θ and g is a scalar source, see Sect. 1.4.3 of the course of G. Falkovich.

Similarly, for a

1b. density field $n(t, \boldsymbol{r})$

(e.g. of a dye), one has the partial differential equation

$$\partial_t n + \boldsymbol{\nabla} \cdot (n\boldsymbol{v}) - \kappa \boldsymbol{\nabla}^2 n = h \tag{2.7}$$

that is different from that for the scalar field only if $\boldsymbol{\nabla} \cdot \boldsymbol{v} \neq 0$, i.e. in the presence of compressibility. Here h is a density source.

For the passive transport of a

2. phase-space density $n(t, \boldsymbol{r}, \boldsymbol{u})$

(e.g. of an aerosol suspension of inertial particles), one can write

$$\partial_t n + (\boldsymbol{u} \cdot \boldsymbol{\nabla_r})n + \boldsymbol{\nabla_u} \cdot (n\frac{\boldsymbol{u} - \boldsymbol{v}}{\tau}) - \kappa \boldsymbol{\nabla}^2 n = h \,,$$

where now h is a source of phase-space density.

Finally, the passive transport of the

3. magnetic field $\boldsymbol{B}(t, \mathrm{r})$

satisfying $\boldsymbol{\nabla} \cdot \boldsymbol{B} = 0$ is governed by the evolution equation

$$\partial_t \boldsymbol{B} + (\boldsymbol{v} \cdot \boldsymbol{\nabla})\boldsymbol{B} + (\boldsymbol{\nabla} \cdot \boldsymbol{v})\boldsymbol{B} - (\boldsymbol{B} \cdot \boldsymbol{\nabla})\boldsymbol{v} - \kappa \boldsymbol{\nabla}^2 \boldsymbol{B} = \boldsymbol{G} \tag{2.8}$$

where κ is the magnetic diffusivity and $\boldsymbol{G}$ is a source term.

What "weak intensity" means in the assumptions leading to the above equation depends on the physical situation (e.g. quite small polymer admixtures may modify considerably the flow [Gro01]). Chemical or biological reactions between the additives, like the ones modeled in J. Cardy's lectures, may also modify their evolution already at weak concentrations.

2.2.2.3 Relation between transport of particles and fields

There is a close relation between the transport of localized objects (particles, polymers) and the passive advection of fields governed by the equations listed above.

Let us first drop the white noise terms in the particle equations and the diffusion and source terms in the field equations. Consider the trajectory

$R(t; t_0, r)$ of the Lagrangian particle that at time t_0 passes through r. If $\partial_t\theta + (v \cdot \nabla)\theta = 0$, then

$$\tfrac{\mathrm{d}}{\mathrm{d}t}\theta(t, R(t; t_0, r)) = (\partial_t\theta)(t, R(t; t_0, r)) + \tfrac{\mathrm{d}R(t; t_0, r)}{\mathrm{d}t} \cdot \nabla\theta(t, R(t; t_0, r))$$

$$= \left[\partial_t\theta + (v \cdot \nabla)\theta\right](t, R(t; t_0, r)) = 0$$

so that the scalar field stays constant along the Lagrangian trajectory. It follows that

$$\theta(t, r) = \theta(t_0, R(t_0; t, r)) = \int \delta(r_0 - R(t_0; t, r))\,\theta(t_0, r_0)\,\mathrm{d}r_0 \qquad (2.9)$$

which determines $\theta(t, r)$ in terms of the time t_0 value of this field. Note that the forward scalar evolution is given in terms of the backward Lagrangian flow.

On the other hand, the forward evolution of the density field satisfying the advection equation $\partial_t n + \nabla \cdot (nv) = 0$ is given in terms of the forward Lagrangian flow:

$$n(t, r) = \int \delta(r - R(t; t_0, r_0))\,n(t_0, r_0)\,\mathrm{d}r_0\,. \qquad (2.10)$$

To see that eq. (2.10) implies the equation for n, consider the integral

$$\int f(r)\,\partial_t n(t, r)\,\mathrm{d}r = \frac{\mathrm{d}}{\mathrm{d}t}\int f(r)\,n(t, r)\,\mathrm{d}r$$

$$= \frac{\mathrm{d}}{\mathrm{d}t}\int f(r)\,\delta(r - R(t; t_0, r_0))\,n(t_0, r_0)\,\mathrm{d}r_0\,\mathrm{d}r$$

$$= \frac{\mathrm{d}}{\mathrm{d}t}\int f(R(t; t_0, r_0))\,n(t_0, r_0)\,\mathrm{d}r_0\,.$$

With the use of the equation of motion of the Lagrangian particle, the last expression may be rewritten as

$$\int [v(t, R(t; t_0, r_0)) \cdot \nabla f(R(t; t_0, r_0))]\,n(t_0, r_0)\,\mathrm{d}r_0$$

$$= \int [v(t, r) \cdot \nabla f(t, r)]\,\delta(r - R(t; t_0, r_0))\,n(t_0, r_0)\,\mathrm{d}r\,\mathrm{d}r_0$$

$$= \int [v(t, r) \cdot \nabla f(r)]\,n(t, r)\,\mathrm{d}r = -\int f(r)\nabla \cdot [n(t, r)v(t, r)]\,\mathrm{d}r$$

and the desired evolution equation follows by using the arbitrariness of f.

To make the relation between the evolutions of scalar and density fields easier to compare, let us consider a matrix $W(t; t_0, \boldsymbol{r}_0)$ with the entries

$$W^i{}_j(t; t_0, \boldsymbol{r}_0) = \frac{\partial R^i(t; t_0, \boldsymbol{r}_0)}{\partial r_0^j}. \tag{2.11}$$

The matrix $W(t; t_0, \boldsymbol{r}_0)$ propagates separations between two infinitesimally close Lagrangian particle trajectories:

$$\boldsymbol{R}(t; t_0, \boldsymbol{r}_0 + \delta \boldsymbol{r}_0) - \boldsymbol{R}(t; t_0, \boldsymbol{r}_0) = W(t; t_0, \boldsymbol{r}_0) \, \delta \boldsymbol{r}_0.$$

Note that the relation $\boldsymbol{R}(t; t_0, \boldsymbol{R}(t_0; t, \boldsymbol{r})) = \boldsymbol{r}$, combined with the chain rule, implies that

$$W(t_0; t, \boldsymbol{r}) = W(t; t_0, \boldsymbol{r}_0)^{-1} \tag{2.12}$$

if $\boldsymbol{r} = \boldsymbol{R}(t; t_0, \boldsymbol{r}_0)$. The identity

$$
\begin{aligned}
\delta(\boldsymbol{r} - \boldsymbol{R}(t; t_0, \boldsymbol{r}_0)) &= \det W(t; t_0, \boldsymbol{r}_0)^{-1} \, \delta(\boldsymbol{r}_0 - \boldsymbol{R}(t_0; t, \boldsymbol{r})) \\
&= \det W(t_0; t, \boldsymbol{r}) \, \delta(\boldsymbol{r}_0 - \boldsymbol{R}(t_0; t, \boldsymbol{r}))
\end{aligned}
$$

permits to rewrite the density field evolution (2.10) in terms of the backward Lagrangian flow as

$$n(t, \boldsymbol{r}) = \det W(t_0; t, \boldsymbol{r}) \, n(t_0, \boldsymbol{R}(t_0; t, \boldsymbol{r})). \tag{2.13}$$

Comparison with eq. (2.9) makes explicit the coincidence between the scalar and density field evolutions in incompressible flows. In compressible flows, however, the density does change along the Lagrangian trajectories in a way inversely proportionally to the space contraction factor $\det W(t; t_0, \boldsymbol{r}_0)$.

To discuss the relation between the polymer motion and the magnetic field evolution, that may seem surprising at the first sight, let us consider the polymer equations of motion (2.4) and (2.5) without noise terms. The second of these equations is solved by setting

$$\boldsymbol{B}(t) = e^{-\alpha(t-t_0)} \, W(t; t_0, \boldsymbol{r}_0) \, \boldsymbol{B}(t_0)$$

since

$$\frac{\mathrm{d}}{\mathrm{d}t} W^i{}_j(t) = \nabla_k v^i(t, \boldsymbol{R}(t)) \, W^k{}_j(t). \tag{2.14}$$

The role of the elasticity coefficient α is clearly visible in this solution. It suppresses the growth of the size of the end-to-end vector with respect to that of an infinitesimal vector frozen into the flow. On the other hand, if

$$\boldsymbol{B}(t, \boldsymbol{r}) = \int \delta(\boldsymbol{r} - \boldsymbol{R}(t; t_0, \boldsymbol{r}_0)) \, W(t; t_0, \boldsymbol{r}_0) \, \boldsymbol{B}(t_0, \boldsymbol{r}_0) \, \mathrm{d}\boldsymbol{r}_0 \tag{2.15}$$

then

$$\partial_t \boldsymbol{B} + (\boldsymbol{v} \cdot \boldsymbol{\nabla})\boldsymbol{B} + (\boldsymbol{\nabla} \cdot \boldsymbol{v})\boldsymbol{B} - (\boldsymbol{B} \cdot \boldsymbol{\nabla})\boldsymbol{v} = 0 \, ,$$

which is the evolution equation of the advected magnetic field for the vanishing magnetic diffusivity and source. The proof goes as for the density $n(t, \boldsymbol{r})$. Note that $W(t; t_0, \boldsymbol{r}_0)$ generates the last term in the equation. Besides, $\boldsymbol{\nabla} \cdot \boldsymbol{B}$ satisfies the same linear homogeneous equation as $n(t, \boldsymbol{r})$ so that if $\boldsymbol{B}(t_0, \boldsymbol{r})$ is divergence-free so is $\boldsymbol{B}(t, \boldsymbol{r})$ given by (2.15).

Finally, for the phase-space motion of an inertial particle $\boldsymbol{R}(t; t_0, \boldsymbol{r}_0, \boldsymbol{u}_0)$ and $\boldsymbol{V}(t; t_0, \boldsymbol{r}_0, \boldsymbol{u}_0)$ governed by eqs. (2.3) without noise terms, one shows the same way as for $n(t, \boldsymbol{r})$ that

$$n(t, \boldsymbol{r}, \boldsymbol{u}) = \int \delta(\boldsymbol{r} - \boldsymbol{R}(t; t_0, \boldsymbol{r}_0, \boldsymbol{u}_0)) \, \delta(\boldsymbol{u} - \boldsymbol{V}(t; t_0, \boldsymbol{r}_0, \boldsymbol{u}_0))$$
$$\cdot \, n(t_0, \boldsymbol{r}_0, \boldsymbol{u}_0) \, \mathrm{d}\boldsymbol{r}_0 \, \mathrm{d}\boldsymbol{u}_0$$

solves the evolution equation

$$\partial_t n + (\boldsymbol{u} \cdot \boldsymbol{\nabla}_{\boldsymbol{r}})n + \boldsymbol{\nabla}_{\boldsymbol{u}} \cdot (n \frac{\boldsymbol{u} - \boldsymbol{v}}{\tau}) = 0$$

of the phase-space density.

The source terms in the field advection equations are taken into account as non-homogeneous terms in any linear equation, i.e. by the variation of the constant method. For example, for the scalar field,

$$\theta(t, \boldsymbol{r}) = \theta(t_0, \boldsymbol{R}(t_0; t, \boldsymbol{r})) + \int_{t_0}^{t} g(s, \boldsymbol{R}(s; t, \boldsymbol{r})) \, \mathrm{d}s \qquad (2.16)$$

solves the transport equation $\partial_t \theta + \boldsymbol{v} \cdot \boldsymbol{\nabla}\theta = g$, compare to Eq. (1.9) in the course of G. Falkovich.

In the presence of the noise, the same expressions for the fields as before, averaged over the noise, solve the field equations with the diffusion terms. For example

$$\theta(t, \boldsymbol{r}) = \overline{\theta(t_0, \boldsymbol{R}(t_0; t, \boldsymbol{r}|\boldsymbol{\eta}))} + \int_{t_0}^{t} \overline{g(s, \boldsymbol{R}(s; t, \boldsymbol{r}|\boldsymbol{\eta}))} \, \mathrm{d}s \, ,$$

solves the advection-diffusion equation (2.6), see Problem 1.3 below. Here, $\boldsymbol{R}(t; t_0, \boldsymbol{r}_0|\boldsymbol{\eta})$ is the solution of eq. (2.2) passing through $\boldsymbol{r}_0$ at time t_0 and the over-line denotes the average over the white noise $\boldsymbol{\eta}$.

Conclusion. *The dynamics of passively advected fields reflects the dynamics of particles transported by the flow. The latter will be studied below by (random) dynamical systems methods. We shall subsequently examine which characteristics of particle dynamics bear on which properties of field advection.*

2.2.3 Problems

2.1   To recall the basic formalism of stochastic differential equations (SDEs) [Oks03, Ris89], consider eq. (2.2) for the noisy tracer particle trajectory $\boldsymbol{R}(t)$ in a given smooth d-dimensional velocity field. In the SDE notation, it is rewritten as an equation for differentials:

$$\mathrm{d}\boldsymbol{R} \;=\; \boldsymbol{v}(t, \boldsymbol{R})\,\mathrm{d}t \;+\; \sqrt{2\kappa}\,\mathrm{d}\boldsymbol{\beta}(t)\,, \qquad (2.17)$$

where $\boldsymbol{\beta}(t)$ is the d dimensional Brownian motion with $\mathrm{d}\boldsymbol{\beta}(t) = \boldsymbol{\eta}(t)\,\mathrm{d}t$. The process $\boldsymbol{\eta}(t)$ is the standard white noise: a Gaussian (generalized) process with mean zero and covariance

$$\overline{\eta^i(t)\,\eta^j(t')} \;=\; \delta^{ij}\,\delta(t - t')\,.$$

Usually, SDEs require a choice of convention, like that of Itô or that of Stratonovich. Show that both conventions coincide for eq. (2.17) but, for any regular function f, the composed process $f(\boldsymbol{R}(t))$ satisfies the Itô SDE

$$\mathrm{d}f(\boldsymbol{R}) = (\boldsymbol{\nabla} f)(\boldsymbol{R}) \cdot [\boldsymbol{v}(t, \boldsymbol{R})\,\mathrm{d}t + \sqrt{2\kappa}\,\mathrm{d}\boldsymbol{\beta}(t)] + \kappa\,(\boldsymbol{\nabla}^2 f)(\boldsymbol{R})\,\mathrm{d}t \quad (2.18)$$

and the Stratonovich one without the last (Itô) term:

$$\mathrm{d}f(\boldsymbol{R}) \;=\; (\boldsymbol{\nabla} f)(\boldsymbol{R}) \cdot [\boldsymbol{v}(t, \boldsymbol{R})\,\mathrm{d}t + \sqrt{2\kappa} \circ \mathrm{d}\boldsymbol{\beta}(t)] \qquad (2.19)$$

(adding a circle is the standard notation for the Stratonovich convention).

2.2   Show that if $\boldsymbol{R}(t; t_0, \boldsymbol{r}_0)$ solves the SDE (2.17) with the initial condition $\boldsymbol{R}(t_0; t_0, \boldsymbol{r}_0) = \boldsymbol{r}_0$ then

$$n(t, \boldsymbol{r}) \;=\; \int \overline{\delta(\boldsymbol{r} - \boldsymbol{R}(t; t_0, \boldsymbol{r}_0))}\; n(t_0, \boldsymbol{r}_0)\,\mathrm{d}\boldsymbol{r}_0 \qquad (2.20)$$

solves the density transport equation (2.7) with vanishing source h.

2.3   With the same notations, show that the solution of the scalar transport equation (2.6) with vanishing source g satisfies the relation

$$\theta(t, \boldsymbol{r}) = \int \overline{\delta(\boldsymbol{r}_0 - \boldsymbol{R}(t_0; t, \boldsymbol{r}))}\; \theta(t_0, \boldsymbol{r}_0)\,\mathrm{d}\boldsymbol{r}_0 = \overline{\theta(t_0, \boldsymbol{R}(t_0; t; \boldsymbol{r}))} \quad (2.21)$$

for $t \geq t_0$.

2.4   For $W^i{}_j(t; t_0, \boldsymbol{r}_0)$ given by eq. (2.11), show that

$$\boldsymbol{B}(t, \boldsymbol{r}) \;=\; \int \overline{\delta(\boldsymbol{r} - \boldsymbol{R}(t; t_0, \boldsymbol{r}_0)) \, W(t; t_0, \boldsymbol{r}_0)} \, \boldsymbol{B}(t_0, \boldsymbol{r}_0) \, \mathrm{d}\boldsymbol{r}_0 \quad (2.22)$$

solves eq. (2.8) for the magnetic field transport without source $\boldsymbol{G}$ for $t \geq t_0$.

2.3 Lecture 2. Multiplicative ergodic theory

This is a lecture devoted to certain general aspects of differentiable dynamical systems. For concreteness, we shall consider the dynamics

$$\frac{\mathrm{d}\boldsymbol{R}}{\mathrm{d}t} = \boldsymbol{v}(t, \boldsymbol{R}) \quad (2.23)$$

of Lagrangian particles carried by the flow in a bounded volume V. In general, we shall deal with random velocity fields $\boldsymbol{v}(t, \boldsymbol{r})$ defined on some probability space $\mathscr{V}$ that we shall take as the very space of the velocity realizations. Below, we shall denote by $\langle \, \cdot \, \rangle$ the expectation with respect to the velocity ensemble. The deterministic case will be also covered by taking space $\mathscr{V}$ composed of one point. Since a general dynamical system has a form similar to eq. (2.23), most of the considerations that follow will also apply, *mutatis mutandis*, to other particle transport equations, although the details of the motion may depend strongly on the case considered.

2.3.1 *Natural measures*

Suppose that we seed Lagrangian particles with constant normalized density $n_{t_0}(t_0, \boldsymbol{r}_0) = |V|^{-1}$ at some past time. This density will evolve according to the relation

$$n_{t_0}(t, \boldsymbol{r}) = \int \delta(\boldsymbol{r} - \boldsymbol{R}(t; t_0, \boldsymbol{r}_0)) \, \frac{\mathrm{d}\boldsymbol{r}_0}{|V|} \,.$$

In the case of an incompressible flow, i.e. for divergence-free velocities with $\boldsymbol{\nabla} \cdot \boldsymbol{v} \equiv 0$, the particle density will stay constant with $n_{t_0}(t, \boldsymbol{r}) = |V|^{-1}$ at all times. If the flow is compressible, however, then particles will cluster during the evolution, developing preferential concentrations and, typically, $n_{t_0}(t, \boldsymbol{r})$ will become rougher and rougher with time. The same phenomenon occurs in the phase space for inertial particles even in incompressible velocities due to the presence of the Stokes friction.

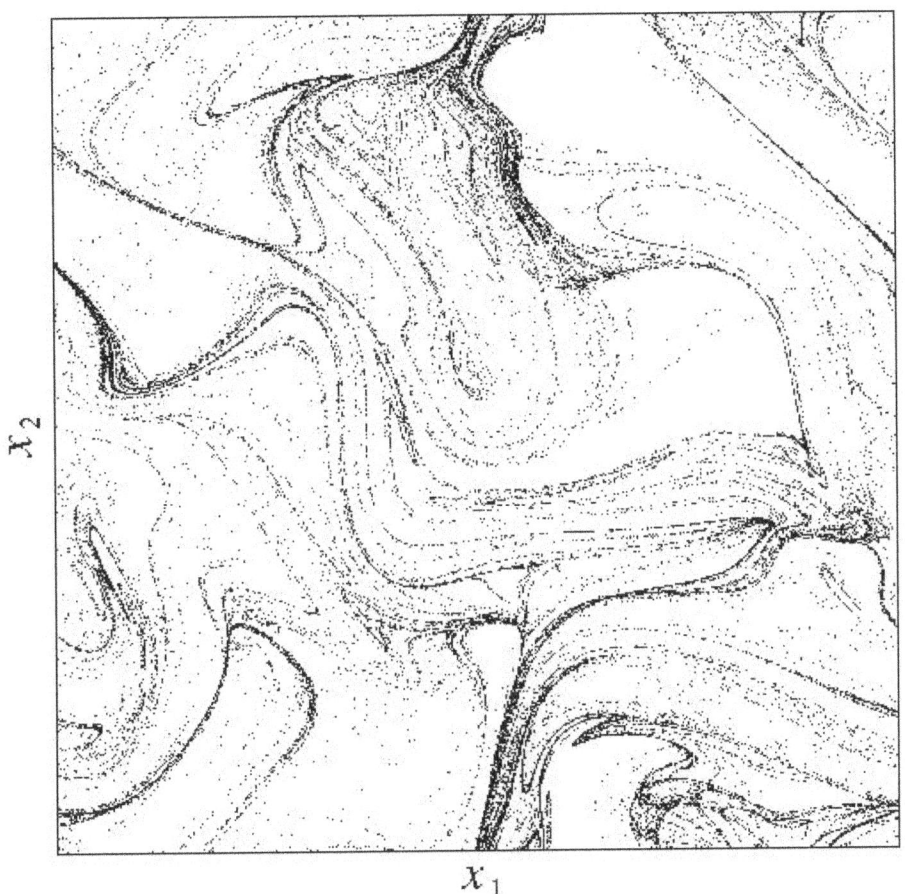

Fig. 2.1. Inertial particles, Figure 6b from [Bec05]

Definition. *The probability measures $n(t, \mathrm{d}r|v)$ on the volume V are called* **natural measures** *if*

$$\int f(r)\, n(t, \mathrm{d}r|v) = \lim_{t_0 \to -\infty} \frac{1}{t - t_0} \int_{t_0}^{t} \mathrm{d}s \int f(r)\, n_{t_0}(s, r|v)\, \mathrm{d}r$$

for all continuous functions f and almost all velocity realizations v.

In particular, $\int f(r)\, n(t, \mathrm{d}r|v) = \lim_{t_0 \to -\infty} \int f(r)\, n_{t_0}(t, r|v)\, \mathrm{d}r$ if the latter limits exist. In that case, natural measures are just the time t distributions of particles seeded with uniform density at time $-\infty$. For compressible flows, natural measures $n(t, \mathrm{d}r|v)$ are concentrated on a random time-dependent attractor. They have the following properties:

$$n(t, \mathrm{d}\boldsymbol{r}|\boldsymbol{v}) = \left(\int n(s, \mathrm{d}\boldsymbol{r}'|\boldsymbol{v})\, \delta(\boldsymbol{r} - \boldsymbol{R}(t; s, \boldsymbol{r}')) \right) \mathrm{d}\boldsymbol{r} \,, \qquad (2.24)$$

$$n(t + \tau, \mathrm{d}\boldsymbol{r}|\boldsymbol{v}) = n(t, \mathrm{d}\boldsymbol{r}|\boldsymbol{v}_\tau) \,, \quad \text{where} \quad \boldsymbol{v}_\tau(t, \boldsymbol{r}) = \boldsymbol{v}(t + \tau, \boldsymbol{r}) \,. \quad (2.25)$$

If we suppose that the velocity ensemble is stationary, i.e. that the time-translated velocity fields $\boldsymbol{v}_\tau$ have the same distribution as $\boldsymbol{v}$, then the natural measures have the same law at different times t. In the stationary case, we shall denote $n(0, \mathrm{d}\boldsymbol{r}|\boldsymbol{v}) \equiv n(\mathrm{d}\boldsymbol{r}|\boldsymbol{v})$ and shall often drop the $\boldsymbol{v}$-dependence from the notation.

Let us consider the combined dynamics on the product $V \times \mathscr{V}$ of the flow volume and the space of velocity realizations:

$$(\boldsymbol{r}, \boldsymbol{v}) \;\xmapsto{\;\Phi_t\;}\; (\boldsymbol{R}(t; \boldsymbol{r}|\boldsymbol{v}), \boldsymbol{v}_t) \,, \qquad (2.26)$$

where we abbreviated $\boldsymbol{R}(t; 0, \boldsymbol{r}|\boldsymbol{v}) \equiv \boldsymbol{R}(t; \boldsymbol{r}|\boldsymbol{v})$. The transformations Φ_t form a one-parameter group:

$$\Phi_t \circ \Phi_s \;=\; \Phi_{t+s} \,, \qquad (2.27)$$

see Problem 2.1 below. It will be interesting to look for probability measures on the space $V \times \mathscr{V}$ that are left invariant during the evolution.

Definition. *A measure $N(\mathrm{d}\boldsymbol{r}, \mathrm{d}\boldsymbol{v})$ on the product space $V \times \mathscr{V}$ is called an **invariant measure** of the random dynamical system (2.23) if it is preserved by the combined dynamics Φ_t and if it projects to $\mathscr{V}$ to the probability measure of the velocity ensemble.*

The first condition means that

$$\int F(\boldsymbol{R}(t; 0, \boldsymbol{r}|\boldsymbol{v}), \boldsymbol{v}_t)\, N(\mathrm{d}\boldsymbol{r}, \mathrm{d}\boldsymbol{v}) = \int F(\boldsymbol{r}, \boldsymbol{v})\, N(\mathrm{d}\boldsymbol{r}, \mathrm{d}\boldsymbol{v}) \qquad (2.28)$$

for all (measurable) functions F on $V \times \mathscr{V}$. The second one signifies that for F independent of $\boldsymbol{r}$,

$$\int F(\boldsymbol{v})\, N(\mathrm{d}\boldsymbol{r}, \mathrm{d}\boldsymbol{v}) = \langle F(\boldsymbol{v}) \rangle \qquad (2.29)$$

(recall that $\langle \,\cdot\, \rangle$ denotes the velocity ensemble average).

Remark. *It follows from the second condition that invariant measures are probability measures and that they may exist only if the velocity ensemble is stationary.*

In the stationary case, one may synthesize an invariant measure on $V \times \mathscr{V}$ out of the time zero natural measures $n(\mathrm{d}\boldsymbol{r}|\boldsymbol{v})$.

Proposition. *In a stationary velocity ensemble, the relation*

$$\int_{V \times \mathscr{V}} F(\boldsymbol{r}, \boldsymbol{v}) \, N(\mathrm{d}\boldsymbol{r}, \mathrm{d}\boldsymbol{v}) = \left\langle \int_V F(\boldsymbol{r}, \boldsymbol{v}) \, n(\mathrm{d}\boldsymbol{r}|\boldsymbol{v}) \right\rangle$$

defines an invariant measure $N(\mathrm{d}\boldsymbol{r}, \mathrm{d}\boldsymbol{v})$ *of the random dynamical system* (2.23).

Proof The property (2.29) is obvious. To prove the property (2.28), note that

$$\left\langle \int_V F(\boldsymbol{r}, \boldsymbol{v}) \, n(\mathrm{d}\boldsymbol{r}|\boldsymbol{v}) \right\rangle = \left\langle \int_V F(\boldsymbol{r}, \boldsymbol{v}_t) \, n(\mathrm{d}\boldsymbol{r}|\boldsymbol{v}_t) \right\rangle = \left\langle \int_V F(\boldsymbol{r}, \boldsymbol{v}_t) \, n(t, \mathrm{d}\boldsymbol{r}|\boldsymbol{v}) \right\rangle,$$

where we have used the stationarity of the velocity ensemble and the relation (2.25). Next, employing the relation (2.24), we obtain

$$\left\langle \int_V F(\boldsymbol{r}, \boldsymbol{v}_t) \, n(t, \mathrm{d}\boldsymbol{r}|\boldsymbol{v}) \right\rangle = \left\langle \int_{V \times V} F(\boldsymbol{r}, \boldsymbol{v}_t) \, n(\mathrm{d}\boldsymbol{r}'|\boldsymbol{v}) \, \delta(\boldsymbol{r} - \boldsymbol{R}(t; \boldsymbol{r}'|\boldsymbol{v})) \, \mathrm{d}\boldsymbol{r} \right\rangle$$
$$= \left\langle \int_V F(\boldsymbol{R}(t; \boldsymbol{r}'|\boldsymbol{v}), \boldsymbol{v}_t) \, n(\mathrm{d}\boldsymbol{r}'|\boldsymbol{v}) \right\rangle = \int_{V \times \mathscr{V}} F(\boldsymbol{R}(t; \boldsymbol{r}|\boldsymbol{v}), \boldsymbol{v}_t) \, N(\mathrm{d}\boldsymbol{r}, \mathrm{d}\boldsymbol{v}).$$

$\square$

We shall call an invariant measure of the random dynamical system (2.23) obtained this way the **natural invariant measure** and we shall suppose that it is **ergodic**, i.e. that (measurable) functions invariant under the dynamics (2.26) are necessarily N-almost everywhere constant. The natural invariant measure turns functions on $V \times \mathscr{V}$ into random variables and permits to do statistics of their values.

2.3.2 Tangent flow

Recall that the matrix with element $W^i{}_j(t; t_0, \boldsymbol{r}|\boldsymbol{v}) = \frac{\partial R^i(t; t_0, \boldsymbol{r}|\boldsymbol{v})}{\partial r^j}$ propagates infinitesimal separations $\delta\boldsymbol{R}$ between Lagrangian trajectories. We shall abbreviate $W(t; 0, \boldsymbol{r}|\boldsymbol{v}) \equiv W(t; \boldsymbol{r}|\boldsymbol{v})$, dropping also the $\boldsymbol{v}$-dependence from the notation whenever it is not necessary. Every real $d \times d$ matrix W may be decomposed as

$$W = \mathcal{O}' \begin{pmatrix} e^{\rho_1} & & \\ & \ddots & \\ & & e^{\rho_d} \end{pmatrix} \mathcal{O} \qquad \text{with } \mathcal{O}, \mathcal{O}' \in O(d), \tag{2.30}$$

where $\rho_1 \geq \ldots \geq \rho_d \geq -\infty$. To achieve such a decomposition, one diagonalizes $W^T W = \mathcal{O}^T \begin{pmatrix} e^{2\rho_1} & \\ & \ddots \\ & & e^{2\rho_d} \end{pmatrix} \mathcal{O}$ and writes a polar decomposition $W = \mathcal{O}''(W^T W)^{1/2}$ taking $\mathcal{O}' = \mathcal{O}'' \mathcal{O}^T$.

Remark.

1. $\mathcal{O}, \mathcal{O}'$ *are not unique. For example, one may multiply $\mathcal{O}$ from the left and $\mathcal{O}'$ from the right by the same diagonal matrix with entries ± 1. On the other hand, $\rho_1 \geq \ldots \geq \rho_d$ are unique since $e^{2\rho_1} \geq \ldots \geq e^{2\rho_d}$ are the ordered eigenvalues of $W^T W$.*
2. *If $\det W > 0$ then $\mathcal{O}$ and $\mathcal{O}'$ may be chosen in $SO(d)$.*

Definition. *The functions $\rho_i(t; \boldsymbol{r}|\boldsymbol{v})$ obtained from the matrices $W(t; \boldsymbol{r}|\boldsymbol{v})$ are called* **stretching exponents**. *They will be treated as random processes on the space $V \times \mathscr{V}$ equipped with the natural invariant measure. The formula*

$$P_t(\vec{\rho}) = \int_{V \times \mathscr{V}} \delta(\vec{\rho} - \vec{\rho}(t; \boldsymbol{r}|\boldsymbol{v})) \, N(\mathrm{d}\boldsymbol{r}, \mathrm{d}\boldsymbol{v})$$

defines their time t joint probability density function (PDF) (which could be distributional).

Note that the chain rule (2.12) implies that

$$W(-t; \boldsymbol{R}(t, \boldsymbol{r}|\boldsymbol{v})|\boldsymbol{v}_t) = W(t; \boldsymbol{r}|\boldsymbol{v})^{-1}.$$

It follows that the backward and forward stretching exponents are simply related:

$$\vec{\rho}(-t; \boldsymbol{R}(t, \boldsymbol{r}|\boldsymbol{v})|\boldsymbol{v}_t) = -\vec{\tilde{\rho}}(t; \boldsymbol{r}|\boldsymbol{v}), \tag{2.31}$$

where $\vec{\tilde{\rho}} = (\rho_d, \ldots, \rho_1)$ if $\vec{\rho} = (\rho_1, \ldots, \rho_d)$. This induces a direct relation between the PDFs $P_t(\vec{\rho})$ and $P_{-t}(\vec{\rho})$, see Problem 2.2 below.

2.3.3 Stretching exponents at long times

We shall be interested in the long-time asymptotic behavior of the stretching exponents that is a cumulative effect of short-time stretchings contractions and rotations of the separation vectors of infinitesimally close Lagrangian trajectories.

Definition. *We shall call the ratios $\sigma_i(t; \boldsymbol{r}|\boldsymbol{v}) = \frac{\rho_i(t; \boldsymbol{r}|\boldsymbol{v})}{t}$ the* **stretching rates**.

2.3.3.1 Multiplicative Ergodic Theorem

First, there is a very general result that holds under a mild assumption for general ergodic invariant measures of random (or deterministic) dynamical systems, see [Arn03]. Denote $\ln_+ x = \max(0, \ln x)$.

Theorem. (MET) *(Oseledec 1968 [Ose68], Ruelle 1979 [Rue79])*

If $\int \ln_+ \|W(t; \boldsymbol{r}|\boldsymbol{v})\| \, N(\mathrm{d}\boldsymbol{r}, \mathrm{d}\boldsymbol{v}) < \infty$ for some $t > 0$ then there exist constants $\lambda_1 \geq \cdots \geq \lambda_d$ such that for N-almost all $(\boldsymbol{r}, \boldsymbol{v})$,

$$\lim_{t \to +\infty} \sigma_i(t; \boldsymbol{r}|\boldsymbol{v}) = \lambda_i = \lim_{t \to -\infty} \sigma_{d-i+1}(t; \boldsymbol{r}|\boldsymbol{v}). \tag{MET}$$

Remark. *In fact, the MET assertion is stronger. It establishes N-almost sure existence of the limits*

$$\lim_{t \to \pm\infty} \frac{1}{t} \ln W^T W(t; \boldsymbol{r}|\boldsymbol{v}) \equiv \Lambda^{\pm}(\boldsymbol{r}|\boldsymbol{v}).$$

The existence of the long time limits of the stretching rates follows. Their constancy is implied then by the assumed ergodicity of N. Additionally, $\mathcal{O}(t; \boldsymbol{r}|\boldsymbol{v}) \xrightarrow[t \to \pm\infty]{} \mathcal{O}^{\pm}(\boldsymbol{r}|\boldsymbol{v})$ for appropriate choices of matrices $\mathcal{O}$ in the decomposition (2.30) of $W(t; \boldsymbol{r}|\boldsymbol{v})$. On the other hand, the matrices $\mathcal{O}'(t, \boldsymbol{r}|\boldsymbol{v})$ do not converge in general.

Definition. $\lambda_1 \geq \cdots \geq \lambda_d \geq -\infty$ *are called* **Lyapunov exponents**. *The stretching rates themselves are sometimes called* **finite time Lyapunov exponents**.

The first Lyapunov exponent gives the exponential growth rate of the length of a typical separation vector $\delta\boldsymbol{R}$ between two infinitesimally close trajectories:

$$\lambda_1 = \lim_{t \to \infty} \frac{1}{t} \ln \|W(t; \boldsymbol{r}|\boldsymbol{v})\delta\boldsymbol{R}\|.$$

Strict positivity of λ_1 signals a sensitive dependence of dynamical system trajectories on the initial conditions and is taken as a definition of **chaos**.

Similarly, for $1 \leq n \leq d$ and typical $\delta\boldsymbol{R}_1, \ldots, \delta\boldsymbol{R}_n$,

$$\lambda_1 + \cdots + \lambda_n = \lim_{t \to \infty} \frac{1}{t} \ln \|W(t; \boldsymbol{r}|\boldsymbol{v})^{\wedge n} \delta\boldsymbol{R}_1 \wedge \cdots \wedge \delta\boldsymbol{R}_n\|$$

$$= \lim_{t \to \infty} \frac{1}{2t} \ln \det\Big((W(t; \boldsymbol{r}|\boldsymbol{v})\delta\boldsymbol{R}_i) \cdot (W(t; \boldsymbol{r}|\boldsymbol{v})\delta\boldsymbol{R}_j) \Big)_{1 \leq i,j \leq n},$$

i.e. $\lambda_1 + \cdots + \lambda_n$ gives the exponential growth rate of the n-dimensional volume spanned by the separation vectors between $n+1$ typical infinitesimally close trajectories.

2.3.3.2 Multiplicative central limit

If $\lambda_1 > \cdots > \lambda_d$ and the temporal correlations of the tangent process decrease sufficiently fast, then one may expect that the fluctuations of the stretching rates about their limiting values behave so that

$$\vec{r}(t; \boldsymbol{r}|\boldsymbol{v}) = \frac{\vec{\rho}(t; \boldsymbol{r}|\boldsymbol{v}) - t\vec{\lambda}}{\sqrt{t}}$$

becomes at long times a normally distributed vector with covariance C and

$$t^{1/2} P_t(\lambda t + \sqrt{t}\vec{r}) \xrightarrow[t\to\infty]{} \frac{e^{-\frac{1}{2}\vec{r}\cdot C^{-1}\vec{r}}}{\det(2\pi C)^{1/2}} .$$

Such a behavior is not assured by a theorem of generality compared to that of the MET.

2.3.3.3 Multiplicative large deviations

Finally, again under the assumption that the Lyapunov exponents are all different and that temporal correlations decay fast enough, one may expect emergence of a large deviation regime at long times, with

$$P_t(\vec{\rho}) \; \propto \; e^{-tH(\frac{\vec{\rho}}{t})} \tag{LD}$$

for a **rate function** $H \geq 0$ such that $H(\vec{\lambda}) = 0$, as argued in [Bal99], see also [Che06]. More exactly the approximate equality should mean that for $A \subset \Delta = \{\vec{\sigma} \mid \sigma_1 \geq \cdots \geq \sigma_d\}$,

$$\limsup_{t\to\infty} \frac{1}{-t} \ln \int_{tA} P_t(\vec{\rho}) \, d\vec{\rho} \; \leq \; \sup_{\vec{\sigma}\in A} H(\vec{\sigma}) \,,$$

$$\liminf_{t\to\infty} \frac{1}{-t} \ln \int_{tA} P_t(\vec{\rho}) \, d\vec{\rho} \; \geq \; \inf_{\vec{\sigma}\in A} H(\vec{\sigma}) \,.$$

As in the additive case, existence of the large deviation regime with a rate function $H(\vec{\sigma})$ twice differentiable around the minimum implies the central limit behavior with

$$(C^{-1})^{ij} = \frac{\partial^2 H}{\partial\sigma_i \partial\sigma_j}(\vec{\lambda})$$

if the Hessian matrix is strictly positive. Again, existence of the multiplicative large deviation regime is not assured by a general theorem. There are, however, partial results about large deviations for deterministic dynamical systems and for random ones with decorrelated velocities, see e.g. [Gra88, Bax88]. The latter case will be discussed in Lecture 3.

2.3.3.4 Multiplicative fluctuation relations

If the velocity field is time-reversible, i.e. the fields $v(t, r)$ and $-v(-t, r)$ have the same distribution, then the following **multiplicative fluctuation relation** (MFR), a multiplicative version of the Gallavotti-Cohen relation [Gal95], should hold:

$$H(-\vec{\sigma}) = H(\vec{\sigma}) - \sum_{i=1}^{d} \sigma_i . \qquad \text{(MFR)}$$

The quantity $-\sum_{i=1}^{d} \rho_i(t) = s(t)$ is equal to the (phase-)space contraction exponent. It was interpreted in [Rue96, Rue97] as the **entropy production** in the dynamical system and $\sigma(t) = s(t)/t$ as the entropy production rate. Setting $\vec{\sigma} = \vec{\lambda}$ in MFR implies that

$$-\sum_{i=1}^{d} \lambda_i \geq 0 . \qquad (2.32)$$

This inequality, that does not require time reversibility, means that the average rate of space contraction around a Lagrangian trajectory is non-negative. This is not too surprising since the trajectories with stronger contraction around them tend to be given a higher weight w.r.t. the natural invariant measure. With the interpretation of $s(t)$ as the entropy production, the inequality (2.32) takes the form of the Second Law of Thermodynamics. MFR also says that in reversible compressible flows, where the inequality is strict, trajectories around which expansion rather than contraction takes place occur with probability suppressed exponentially for long times.

Let us prove here another, simpler, **transient multiplicative fluctuation relation** (TMFR). It is a multiplicative version of the Evans-Searles relation [Eva94] and it was formulated in the Lagrangian flow context in [Bal01], see also [Fal01]. It deals with a modified joint PDF of the time t stretching exponents defined by

$$\tilde{P}_t(\vec{\rho}) = \left\langle \int \delta(\vec{\rho} - \vec{\rho}(t; r|v)) \frac{\mathrm{d}r}{|V|} \right\rangle . \qquad (2.33)$$

Note that $\tilde{P}_t$ employs $\frac{\mathrm{d}r}{|V|}$ rather than the natural measures $n(\mathrm{d}r|v)$ used in P_t in order to average over the initial space points r.

Proposition. *If the velocity field is time-reversible then*

$$\tilde{P}_t(-\vec{\rho}) = \tilde{P}_t(\vec{\rho}) \, e^{\sum_i \rho_i} \qquad \text{(TMFR)}$$

Proof This is a somewhat tautological relation that follows by a simple change of variables. First, using the definition (2.33) and the stationarity of the velocity ensemble, we obtain

$$\tilde{P}_{-t}(-\vec{\rho}) = \left\langle \int \delta(-\vec{\rho} - \vec{\rho}(-t; \boldsymbol{R}|\boldsymbol{v})) \frac{\mathrm{d}\boldsymbol{R}}{|V|} \right\rangle$$

$$= \left\langle \int \delta(\vec{\rho} + \vec{\rho}(-t; \boldsymbol{R}|\boldsymbol{v}_t)) \frac{\mathrm{d}\boldsymbol{R}}{|V|} \right\rangle.$$

Upon the substitution

$$\boldsymbol{R} = \boldsymbol{R}(t; \boldsymbol{r}|\boldsymbol{v}), \qquad \frac{\partial(\boldsymbol{R})}{\partial(\boldsymbol{r})} = \det W(t; \boldsymbol{r}|\boldsymbol{v}) = e^{\sum_{i=1}^{d} \rho_i(t; \boldsymbol{r}|\boldsymbol{v})},$$

this gives

$$\tilde{P}_{-t}(-\vec{\rho}) = \left\langle \int \delta(\vec{\rho} + \vec{\rho}(-t; \boldsymbol{R}(t; \boldsymbol{r}|\boldsymbol{v})|\boldsymbol{v}_t)) \, e^{\sum_{i=1}^{d} \rho_i(t; \boldsymbol{r}|\boldsymbol{v})} \frac{\mathrm{d}\boldsymbol{r}}{|V|} \right\rangle$$

$$= \left\langle \int \delta(\vec{\rho} - \vec{\rho}(t; \boldsymbol{r}|\boldsymbol{v})) \, e^{\sum_{i=1}^{d} \rho_i(t; \boldsymbol{r}|\boldsymbol{v})} \frac{\mathrm{d}\boldsymbol{r}}{|V|} \right\rangle = \tilde{P}_t(\vec{\rho}) \, e^{\sum_{i=1}^{d} \rho_i}, \quad (2.34)$$

where we have used the relation (2.31). Now, if $\boldsymbol{v}$ is time-reversible, then $\tilde{P}_{-t}(\vec{\rho}) = \tilde{P}_t(\vec{\rho})$ and TMFR follows. $\qquad\square$

One expects that at long positive times t,

$$\tilde{P}_t(\vec{\rho}) \approx P_t(\vec{\rho}) \qquad\qquad (2.35)$$

since Lagrangian trajectories stay close to the attractor at late times and could be equivalently sampled with the attractor measure. If the relation (2.35) holds for $\vec{\rho} = O(t)$ and $P_t(\vec{\rho})$ has a large deviation regime then TMFR implies MFR. Note that the relation (2.34), that did not use time reversibility, implies that $\int \tilde{P}_t(\vec{\rho}) \, e^{\sum \rho_i} \, \mathrm{d}\vec{\rho} = 1$ and, by using the Jensen equality, that $\int \sum \rho_i \, \tilde{P}_t(\vec{\rho}) \, \mathrm{d}\vec{\rho} \leq 0$. This is another expression of the fact that in compressible flows there is typically a compression around Lagrangian trajectories.

Remarks.

1. *The difficulty in passing from TMFR to MFR lies in showing that the approximate equation (2.35) holds in a sufficiently strong sense and that P_t exhibits a large deviation regime. Both relations require strong mixing properties of the dynamics. They will hold in the Kraichnan model.*

2. *The original fluctuation relations observed over a decade ago [Eva93] and intensively studied ever since, dealt with fluctuations of the (phase-)space*

contraction exponent $s(t)$. For deterministic uniformly hyperbolic systems, this exponent, sampled with the so called SRB natural invariant measure, see e.g. [You02], has large deviations controlled by a rate function $h(\sigma)$. Gallavotti and Cohen [Gal95] established that

$$h(-\sigma) = h(\sigma) + \sigma. \tag{FR}$$

Recently, this was generalized to random uniformly hyperbolic systems in [Bon06]. Since

$$h(\sigma) = \min_{\sum \sigma_i = \sigma} H(-\vec{\sigma}),$$

MFR implies FR.

The multiplicative large deviations regime is difficult to access in realistic flows. For surface flows, there have been recently some attempts to measure the rate function $H(\vec{\sigma})$ in numerical simulations [Bof06] and to experimentally check the Gallavotti-Cohen fluctuation relation [Ban06]. The surface flows are governed by the horizontal components of velocity and are compressible for incompressible bulk flows.

Conclusion. *In smooth dynamical systems, local stretchings, contractions and rotations along trajectories lead to a cumulative effect that is quantitatively captured by the Lyapounov exponents, and, with more precision, by the rate function $H(\vec{\sigma})$ of large deviations for stretching exponents. In reversible flows, the rate function possesses a symmetry of the Gallavotti-Cohen type.*

2.3.4 Problems

2.1   Show the group property (2.27) of the combined dynamics Φ_t.

2.2   Prove that

$$P_{-t}(\vec{\rho}) = P_t(-\vec{\rho}). \tag{2.36}$$

2.3   Explain why eq. (2.36) and the identity

$$\tilde{P}_{-t}(\vec{\rho}) = \tilde{P}_t(-\vec{\rho}) \, e^{-\sum_i \rho_i},$$

that follows from eq. (2.34), may be consistent with the behavior (2.35).

2.4 Lecture 3.   Kraichnan model

In 1968, Robert H. Kraichnan has proposed to study the turbulent transport in a random Gaussian field of velocities decorrelated in time [Kra68]. The statistics of such velocities is completely determined by the mean $\langle v^i(t, \boldsymbol{r}) \rangle$, that we shall take vanishing, and by the covariance of the form

$$\langle v^i(t, \boldsymbol{r})\, v^j(t', \boldsymbol{r}') \rangle = \delta(t - t')\, D^{ij}(\boldsymbol{r}, \boldsymbol{r}')\,.$$

To mimic turbulent flows far from boundaries, one may assume the following properties of the velocity ensemble:

- homogeneity: $\quad\quad\quad D^{ij}(\boldsymbol{r}, \boldsymbol{r}') = D^{ij}(\boldsymbol{r} - \boldsymbol{r}')\,,$

- isotropy: $\quad\quad\quad\quad\; D^{ij}(\boldsymbol{r}) = \delta^{ij} D'(|\boldsymbol{r}|) + \dfrac{r^i r^j}{r^2} D''(|\boldsymbol{r}|)\,,$

- scaling: $\quad\quad\quad\quad\;\; D^{ij}(\boldsymbol{r}) = \begin{cases} O(\boldsymbol{r}^2) & \text{for } |\boldsymbol{r}| \ll \eta\,, \\ O(|\boldsymbol{r}|^\xi) & \text{for } \eta \ll |\boldsymbol{r}| \ll L\,. \end{cases}$

The range where $D^{ij}(\boldsymbol{r})$ is approximately quadratic is called the **Batchelor regime** and it represents a flow dominated by viscous effects. The range where $D^{ij}(\boldsymbol{r})$ scales with a power ξ between 0 and 2 mimics the **inertial range** of developed turbulence (the Kolmogorov scaling corresponds to $\xi = 4/3$). Although not very realistic, the Kraichnan model has appeared to be extremely rich and it allowed for multiple insights into the intricacies of turbulent transport [Fal01]. The most important of them will be discussed below.

2.4.1 Lagrangian trajectories and eddy diffusion

In the Kraichnan ensemble, velocities are white noise (so distributional) in time and the Lagrangian trajectory equation becomes the stochastic differential equation (SDE)

$$\mathrm{d}\boldsymbol{R} = \mathrm{d}\boldsymbol{w}(t, \boldsymbol{R})\,, \tag{2.37}$$

where $\mathrm{d}\boldsymbol{w}(t, \boldsymbol{R}) = \boldsymbol{v}(t, \boldsymbol{r})\, \mathrm{d}t$ so that $\boldsymbol{w}(t, \boldsymbol{R})$ is the time-integrated velocity. We shall consider this SDE with the Itô convention. It is, however, equal to the one with the Stratonovich convention. This is the subject of Problem 2.1 below. In the sequel, we shall abandon using $\mathrm{d}\boldsymbol{w}(t, \boldsymbol{r})$ in the notation, replacing it by $\boldsymbol{v}(t, \boldsymbol{r})\, \mathrm{d}t$ to keep the formulae closer to the standard ones for velocities regular in time. A function $f(\boldsymbol{R})$ of the process that solves the SDE (2.37) with the Itô convention satisfies the Itô SDE

$$\mathrm{d}f(\boldsymbol{R}) = (\boldsymbol{\nabla} f)(\boldsymbol{R}) \cdot \boldsymbol{v}(t, \boldsymbol{R})\, \mathrm{d}t + \tfrac{1}{2} D^{ij}(0) \nabla_i \nabla_j f(\boldsymbol{R})\, \mathrm{d}t\,. \tag{2.38}$$

The last contribution is the Itô term. Due to the assumed isotropy, $D^{ij}(0) = D_0 \delta^{ij}$ and we obtain from the SDE (2.38) the relation

$$\frac{\mathrm{d}}{\mathrm{d}t}\langle f(\boldsymbol{R})\rangle = \tfrac{1}{2}D_0\langle \boldsymbol{\nabla}^2 f(\boldsymbol{R})\rangle \tag{2.39}$$

for the average values. Let $P(t, \boldsymbol{r}; t_0, \boldsymbol{r}_0) = \langle\delta(\boldsymbol{r} - \boldsymbol{R}(t; t_0, \boldsymbol{r}_0))\rangle$ denote the PDF to find a Lagrangian trajectory at point $\boldsymbol{r}$ at time t if it started at point $\boldsymbol{r}_0$ at time t_0. We have the relation:

$$\langle f(\boldsymbol{R}(t; t_0, \boldsymbol{r}_0))\rangle = \int f(\boldsymbol{r})\, P(t, \boldsymbol{r}; t_0, \boldsymbol{r}_0)\, \mathrm{d}\boldsymbol{r}$$

so that the identity (2.39) implies the heat equation

$$\frac{\mathrm{d}}{\mathrm{d}t} P(t, \boldsymbol{r}; t_0, \boldsymbol{r}_0) = \tfrac{1}{2}D_0 \boldsymbol{\nabla}_r^2 P(t, \boldsymbol{r}; t_0, \boldsymbol{r}_0)\,.$$

Consequently, the PDF $P(t, \boldsymbol{r}; t_0, \boldsymbol{r}_0)$ is as for a diffusing particle

$$P(t, \boldsymbol{r}; t_0, \boldsymbol{r}_0) = e^{\frac{1}{2}|t-t_0|D_0 \boldsymbol{\nabla}^2}(\boldsymbol{r}; \boldsymbol{r}_0) = \frac{1}{(2\pi D_0|t - t_0|)^{d/2}} e^{-\frac{(\boldsymbol{r}-\boldsymbol{r}_0)^2}{2D_0|t-t_0|}}\,.$$

The constant $\tfrac{1}{2}D_0$ is called the **eddy diffusivity**. We infer that, in mean, the Kraichnan turbulence causes diffusion. This effect, called eddy diffusion, holds in more general flows at sufficiently long time and distance scales under quite weak conditions [Tay21]. For the average scalar field, one obtains from eq. (2.9) the relation

$$\langle\theta(t, \boldsymbol{r})\rangle = \left\langle \int \theta(t_0, \boldsymbol{r}_0)\, \delta(\boldsymbol{r}_0 - \boldsymbol{R}(t_0; t, \boldsymbol{r}))\, \mathrm{d}\boldsymbol{r}_0 \right\rangle$$
$$= \int \theta(t_0, \boldsymbol{r}_0)\, e^{\frac{1}{2}|t-t_0|D_0 \boldsymbol{\nabla}^2}(\boldsymbol{r}_0; \boldsymbol{r})\, \mathrm{d}\boldsymbol{r}_0\,,$$

and, for the average density, from eq. (2.10):

$$\langle n(t, \boldsymbol{r})\rangle = \left\langle \int n(t_0, \boldsymbol{r}_0)\, \delta(\boldsymbol{r} - \boldsymbol{R}(t; t_0, \boldsymbol{r}_0))\, \mathrm{d}\boldsymbol{r}_0 \right\rangle$$
$$= \int n(t_0, \boldsymbol{r}_0)\, e^{\frac{1}{2}|t-t_0|D_0 \boldsymbol{\nabla}^2}(\boldsymbol{r}, \boldsymbol{r}_0)\, \mathrm{d}\boldsymbol{r}_0\,.$$

An initial blob of scalar or density diffuses in mean as watched in the laboratory frame. In realistic flows, emergence of such an evolution of advected scalars at large scales, smearing the small-scale details, is called **homogenization**, see [Maj99, Pav06].

2.4.2 Tangent flow in Kraichnan velocities

It will be more interesting to look at the evolution of the blob in a reference frame that moves with the fluid. To this end, let us first study the dynamics of a small (infinitesimal) separation $\delta \boldsymbol{R}$ between Lagrangian trajectories that satisfies the SDE

$$\mathrm{d}\delta \boldsymbol{R} \;=\; (\delta \boldsymbol{R} \cdot \boldsymbol{\nabla}) \boldsymbol{v}(t, \boldsymbol{R}(t))\, \mathrm{d}t \;\equiv\; \Sigma(t)\mathrm{d}t\,\delta \boldsymbol{R}\,, \tag{2.40}$$

where $\Sigma^i{}_j(t) = \nabla_j v^i(t, \boldsymbol{R}(t))$. Again, we shall consider this equation with the Itô convention, but the Stratonovich convention would give the same result, see Problem 2.2 below. For a function of $\delta \boldsymbol{R}$, the Itô calculus implies that

$$\mathrm{d}f(\delta \boldsymbol{R}) \;=\; (\delta \boldsymbol{R} \cdot \boldsymbol{\nabla}) \boldsymbol{v}(t, \boldsymbol{R}(t))\, \mathrm{d}t \cdot \boldsymbol{\nabla} f(\delta \boldsymbol{R})$$

$$-\; \tfrac{1}{2}\delta R^k \delta R^l \nabla_k \nabla_l D^{ij}(\boldsymbol{0}) \nabla_i \nabla_j f(\delta \boldsymbol{R})\, \mathrm{d}t$$

and hence that

$$\frac{\mathrm{d}}{\mathrm{d}t}\big\langle f(\delta \boldsymbol{R})\big\rangle = \big\langle \tfrac{1}{2}\delta R^k \delta R^l C^{ij}_{kl} \nabla_i \nabla_j f(\delta \boldsymbol{R})\big\rangle\,, \tag{2.41}$$

where $C^{ij}_{kl} = -\nabla_k \nabla_l D^{ij}(\boldsymbol{0})$. The last result is the same as if $\delta \boldsymbol{R}(t)$ satisfied the linear Itô SDE

$$\mathrm{d}\delta \boldsymbol{R} = S(t)\mathrm{d}t\,\delta \boldsymbol{R} \tag{2.42}$$

with the matrix-valued white noise $S(t)$ of vanishing mean and covariance

$$\big\langle S^i{}_k(t)\, S^j{}_l(t')\big\rangle \;=\; \delta(t - t')\, C^{ij}_{kl}\,. \tag{2.43}$$

Here taking the Itô prescription is essential since, although the Itô and Stratonovich conventions give the same result for eq. (2.40) with $\Sigma^i{}_j(t) = \nabla_j v^i(t, \boldsymbol{R}(t))$, they do not agree for eq. (2.42) with $S^i{}_j(t)$ being the white noise distributed as $\nabla_j v^i(t, \boldsymbol{r})$ for fixed $\boldsymbol{r}$, see Problem 2.3 below.

The statistics of the tangent process $W(t; t_0, \boldsymbol{r}_0)$ for fixed $\boldsymbol{r}_0$ may be obtained similarly by solving the linear Itô SDE

$$\mathrm{d}W = S(t)\mathrm{d}t\,W \tag{2.44}$$

with the white noise $S(t)$ instead of $\Sigma(t)$ and $W(t_0) = \mathrm{Id}$. The possibility to drop the dependence on $\boldsymbol{R}(t)$ when we study the statistics of the tangent process with fixed initial point is the great simplification of the Kraichnan model!

Natural measures $n(\mathrm{d}\boldsymbol{r}|\boldsymbol{v})$ seem to exist in the Kraichnan model in a finite volume. In the periodic box with homogeneous velocities, one has to have

$$\langle n(\mathrm{d}\boldsymbol{r}|\boldsymbol{v})\rangle = \frac{\mathrm{d}\boldsymbol{r}}{|V|}.$$

The natural measure $n(\mathrm{d}\boldsymbol{r}|\boldsymbol{v})$ depends on past velocities and, for $t \geq 0$, $W(t, \boldsymbol{r}|\boldsymbol{v})$ depends on future ones. In the Kraichnan model, past and future velocities are independent. It follows then that

$$\left\langle \int_V f(W(t;\boldsymbol{r}|\boldsymbol{v}))\, n(\mathrm{d}\boldsymbol{r}|\boldsymbol{v}) \right\rangle = \int_V \langle f(W(t;\boldsymbol{r}|\boldsymbol{v}))\rangle\langle n(\mathrm{d}\boldsymbol{r}|\boldsymbol{v})\rangle$$
$$= \int_V \langle f(W(t;\boldsymbol{r}|\boldsymbol{v}))\rangle \frac{\mathrm{d}\boldsymbol{r}}{|V|} = \langle f(W(t;\boldsymbol{r}_0|\boldsymbol{v}))\rangle.$$

Corollary.

1. *In the homogeneous Kraichnan model, the statistics of $W(t; \boldsymbol{r}|\boldsymbol{v})$ for $t \geq 0$, with $(\boldsymbol{r}, \boldsymbol{v})$ distributed according to the natural invariant measure $N(\mathrm{d}\boldsymbol{r}, \mathrm{d}\boldsymbol{v})$, coincides with the statistics of the solution $W(t)$ of eq. (2.44) with the matrix-valued white noise $S(t)$ and the initial condition $W(0) = $ Id. Consequently, also the distributions of the corresponding vectors of stretching exponents $\vec{\rho}(t; \boldsymbol{r}|\boldsymbol{v})$ and $\vec{\rho}(t)$ coincide.*

2. *It follows that $P_t(\vec{\rho}) = \tilde{P}_t(\vec{\rho})$ for $t \geq 0$ (recall that $P_t(\vec{\rho})$ was obtained by sampling the initial points with the natural measures and $\tilde{P}_t(\vec{\rho})$ with the uniform one). As the Kraichnan ensemble is time reversible, the transient multiplicative fluctuation relation TMFR proved in Lecture 2 implies that*

$$P_t(-\vec{\rho}) = P_t(\vec{\rho})\, e^{\sum_{i=1}^{d} \rho_i}$$

for all $t \geq 0$.

3. *In the homogeneous Kraichnan model, MFR follows from TMFR if $P_t(\vec{\rho})$ exhibits the large deviation behavior (LD) at long times, see below.*

The assumption of isotropy restricts the possible form of the covariance C^{ij}_{kl},

$$C^{ij}_{kl} = \beta(\delta^i_k \delta^j_l + \delta^i_l \delta^j_k) + \gamma \delta^{ij}\delta_{kl}, \tag{2.45}$$

with the positivity requiring that $\gamma \geq |\beta|$ and $(d+1)\beta + \gamma \geq 0$. Up to normalization, the right hand side is specified by the **compressibility degree**

$$\wp = \frac{\langle (\nabla_i v^i)^2 \rangle}{\langle (\nabla_j v^i)^2 \rangle} = \frac{C^{ij}_{ij}}{C^{ii}_{jj}} = \frac{(d+1)\beta + \gamma}{2\beta + d\gamma}. \tag{2.46}$$

Vanishing $\wp$ corresponds to incompressible velocities, whereas $\wp = 1$ to gradient ones. In general, $0 \leq \wp \leq 1$ and it measures the degree of local compression in the flow.

By Itô calculus, the average of a regular function f of a solution $W(t)$ of eq. (2.44) evolves according to the equation

$$\frac{\mathrm{d}}{\mathrm{d}t}\langle f(W)\rangle = \langle (\mathscr{L}f)(W)\rangle\,,$$

where

$$\mathscr{L} = \tfrac{1}{2} C^{ij}_{kl} W^k{}_m W^l{}_n \frac{\partial}{\partial W^i{}_m} \frac{\partial}{\partial W^j{}_n}\,. \qquad (2.47)$$

Observe that if the process $W(t)$ satisfies the Itô SDE (2.44) then, for orthogonal matrices $\mathcal{O}, \mathcal{O}' \in O(d)$, the process $W'(t) \equiv \mathcal{O}'W(t)\mathcal{O}$ satisfies the Itô SDE

$$\mathrm{d}W' = \mathcal{O}'S(t)\mathcal{O}'^{-1}W'\mathrm{d}t\,, \qquad (2.48)$$

where $S'(t) \equiv \mathcal{O}'S(t)\mathcal{O}'^{-1}$ is the matrix-valued white noise with the same distribution as $S(t)$, by the assumption of isotropy. It follows that for $f'(W) = f(\mathcal{O}'W\mathcal{O})$,

$$(\mathscr{L}f')(W) = (\mathscr{L}f)(\mathcal{O}'W\mathcal{O}) \qquad (2.49)$$

(we could even take $\mathcal{O} \in GL(d)$). Consequently, if a function f is left-right $O(d)$−invariant,

$$f(\mathcal{O}'W\mathcal{O}) = f(W) \qquad \text{for } \mathcal{O}', \mathcal{O} \in O(d),$$

so is $\mathscr{L}f$. The left-right $O(d)$−invariant functions may be identified with functions of the stretching rate vector $\vec{\rho}$ which we shall denote, somewhat abusively, by the same symbol. In order to find $\frac{\mathrm{d}}{\mathrm{d}t}\langle f(\vec{\rho})\rangle$, it is then enough to calculate $\mathscr{L}$ in the action on the corresponding left and right $O(d)$-invariant function on $GL(d)$. This is tedious but straightforward.

Proposition.

1. $\frac{\mathrm{d}}{\mathrm{d}t}\langle f(\vec{\rho})\rangle = \langle (\mathcal{L}f)(\vec{\rho})\rangle$ *for*

$$\begin{aligned}
\mathcal{L} &= \frac{\beta+\gamma}{2}\Big(\sum_i \frac{\partial^2}{\partial\rho_i^2} + \sum_{i\neq j} \coth(\rho_i - \rho_j)\frac{\partial}{\partial\rho_i}\Big) \\
&+ \frac{\beta}{2}\Big(\sum_i \frac{\partial}{\partial\rho_i}\Big)^2 - \frac{(d+1)\beta+\gamma}{2}\sum_i \frac{\partial}{\partial\rho_i}
\end{aligned} \qquad (2.50)$$

2. For $\mathcal{F}(\vec{\rho}) = e^{-\frac{1}{2}\sum_i \rho_i}(\prod_{i<j} \sinh(\rho_i - \rho_j))^{1/2}$,

$$\mathcal{F}\mathcal{L}\mathcal{F}^{-1} = \frac{\beta+\gamma}{2}\Big(\sum_i \frac{\partial^2}{\partial\rho_i^2} + \frac{1}{2}\sum_{i<j}\frac{1}{\sinh^2(\rho_i - \rho_j)}\Big) + \frac{\beta}{2}\Big(\sum_i \frac{\partial}{\partial\rho_i}\Big)^2$$
$$- \text{const.} \equiv -\mathcal{H}_{\text{CSM}}, \qquad (2.51)$$

where $\mathcal{H}_{\text{CSM}}$ is the Calogero-Sutherland-Moser Hamiltonian of d particles on the line.

Proof See Problems 2.4 to 2.6 below.

Since

$$\langle f(\vec{\rho}(t))\rangle = \int f(\vec{\rho})\, P_t(\vec{\rho})\, \mathrm{d}\vec{\rho}$$

if $\vec{\rho}(0) = \vec{0}$, it follows that

$$P_t(\vec{\rho}) = e^{t\mathcal{L}}(\vec{0}; \vec{\rho}) = \lim_{\vec{\rho}_0 \to \vec{0}} \mathcal{F}(\vec{\rho}_0)^{-1} e^{-t\mathcal{H}_{\text{CSM}}}(\vec{\rho}_0; \vec{\rho})\, \mathcal{F}(\vec{\rho})\,.$$

The spectral representation of the heat kernel $e^{-t\mathcal{H}_{\text{CSM}}}$ is explicitly known [Ols83]. Its saddle point calculation gives the large deviation form of $P_t(\vec{\rho})$. The latter may also be found directly [Bal99] by approximating

$$P_t(\vec{\rho}) \simeq e^{t\mathcal{L}_{\text{asym}}}(\vec{0}; \vec{\rho})\,,$$

where $\mathcal{L}_{\text{asym}}$ is obtained from $\mathcal{L}$ by replacing

$$\cosh(\rho_i - \rho_j) \longrightarrow \begin{cases} 1 & i > j\,, \\ -1 & i < j\,. \end{cases}$$

We have

$$\mathcal{L}_{\text{asym}} = \frac{\beta+\gamma}{2}\sum_i \frac{\partial^2}{\partial\rho_i^2} + \frac{\beta}{2}\Big(\sum_i \frac{\partial}{\partial\rho_i}\Big)^2 + \sum_i \Big(\frac{\beta+\gamma}{2}(d-2i) - \frac{\beta d}{2}\Big)\frac{\partial}{\partial\rho_i}\,.$$

Since $\mathcal{L}_{\text{asym}}$ is a second order operator on $\mathbb{R}^d$ with constant coefficients, the kernel $e^{t\mathcal{L}_{\text{asym}}}(\vec{0}; \vec{\rho})$ is Gaussian, up to a normalizing factor. It is easy to check that for all times

$$e^{t\mathcal{L}_{\text{asym}}}(\vec{0}; \vec{\rho}) = \frac{1}{N_t}\, e^{-t\,H(\frac{\vec{\rho}}{t})}$$

with

$$H(\vec{\sigma}) = \frac{1}{2(\beta + \gamma)} \left[\sum_i (\sigma_i - \lambda_i)^2 - \frac{\beta}{(d+1)\beta + \gamma} \left(\sum_i (\sigma_i - \lambda_i) \right)^2 \right], \quad (2.52)$$

$$\lambda_i = \frac{\beta + \gamma}{2}(d - 2i) - \frac{\beta d}{2}, \quad (2.53)$$

$$N_t = \sqrt{(2\pi t)^d (\beta + \gamma)^{d-1}((d+1)\beta + \gamma)}.$$

The constants λ_i are the Lyapunov exponents. Note that they are equally spaced. The chaotic phase with $\lambda_1 > 0$ occurs for $\wp < d/4$ whereas $\lambda_1 < 0$ for $\wp > d/4$. The change of sign of λ_1 induces a change in the transport properties (direct cascade of the the the scalar for $\lambda_1 > 0$ versus inverse cascade for $\lambda_1 < 0$, see [Che98]).

2.4.3 The uses of multiplicative large deviations

Suppose that we put at time zero a blob of density $n_0(\boldsymbol{r})$ into a compressible homogeneous flow. The evolution of the blob density along a fixed Lagrangian trajectory $\boldsymbol{R}(t; \boldsymbol{r}_0)$ starting at time $t_0 = 0$ at point $\boldsymbol{r}_0$ (e.g. at the center of the blob) is given by the relation

$$n(t) \equiv n(t, \boldsymbol{R}(t; \boldsymbol{r}_0)) = \det W(t; \boldsymbol{r}_0)^{-1} n_0(\boldsymbol{r}_0) = e^{-\sum_i \rho_i(t; \boldsymbol{r}_0)} n_0(\boldsymbol{r}_0),$$

see eqs. (2.13) and (2.12). Hence

$$\left\langle n(t)^\alpha \right\rangle = n_0(\boldsymbol{r}_0)^\alpha \int e^{-\alpha \sum \rho_i} \tilde{P}_t(\vec{\rho}) \, d\vec{\rho}$$

$$\approx n_0(\boldsymbol{r}_0)^\alpha \int e^{-\alpha \sum \rho_i - tH(\frac{\vec{\rho}}{t})} \, d\vec{\rho} \propto e^{\gamma_\alpha t}$$

for large t, where $-\gamma_\alpha = \min_{\sigma_1 \geq \cdots \geq \sigma_d} \alpha \sum_i \sigma_i + H(\vec{\sigma})$. For the Kraichnan model,

$$\gamma_\alpha = d[(d+1)\beta + \gamma] \left[\frac{\alpha}{2} + \frac{\alpha^2}{(\beta + \gamma)^2} \right].$$

Note that $\gamma_\alpha = 0$ for $\wp = 0$ but γ_α is positive for $\wp > 0$ and $\alpha > 0$ so that, in compressible flows, the positive moments of the density along a Lagrangian trajectory grow exponentially for long times. Their growth rates provide a quantitative measure of clustering of Lagrangian particles in the presence of compressibility.

Among other quantities that may be expressed in terms of the rate function $H(\vec{\sigma})$ of large deviations for stretching exponents there are:

- the decay of scalar moments in the presence of diffusivity [Bal99],

- the multifractal dimensions of the natural invariant measure [Bec04],

- the evolution of moments of polymer end-to-end length $|\boldsymbol{B}|$ in polymer solutions [Che00, Bal00].

Remark. *The isotropy of the covariance C^{ij}_{kl} of the matrix white noise $S(t)$ is a strong condition and is generically broken by putting the Kraichnan model into a periodic box, leading to non-Gaussian multiplicative large deviations [Che06].*

Conclusion. *In many transport problems in the Batchelor regime of turbulence it is not enough to know the Lyapunov exponents of the flow. Instead, it is the knowledge of the multiplicative large deviations rate function $H(\vec{\sigma})$ that is required. The latter is analytically accessible in the homogeneous isotropic Kraichnan model.*

2.4.4 Problems

2.1 Show that the SDE (2.37) for the Lagrangian trajectories in the Kraichnan ensemble of homogeneous isotropic velocities $\boldsymbol{v}$ is the same with the Itô and and with the Stratonovich conventions.

2.2 Show that in the same velocity ensemble, the SDE (2.40) for the infinitesimal separation of the Lagrangian trajectories is the same with the Itô and the Stratonovich conventions.

2.3 Show that for the SDE (2.42), where $S(t)$ is the matrix-valued white noise with covariance (2.43), the Itô and the Stratonovich conventions do not coincide, in general.

2.4 Let $\mathscr{E}_i^{\ j}$ be the generators of the left action of the Lie algebra $gl(d)$ on functions on $GL(d)$:

$$(\mathscr{E}_i^{\ j} f)(W) = \left.\frac{\mathrm{d}}{\mathrm{d}\epsilon}\right|_0 f(e^{-\epsilon E_i^{\ j}} W)$$

for $E_i^{\ j}$ denoting the matrices with 1 on the intersection of the i-th row and j-th column and zeros elsewhere. Let $\mathscr{D}$ denote the generator of dilations:

$$(\mathscr{D} f)(W) = \left.\frac{\mathrm{d}}{\mathrm{d}\epsilon}\right|_0 f(e^{\epsilon} W).$$

Prove that

$$\mathscr{L} = \frac{\beta}{2} \sum_{i,j} \mathcal{E}_i{}^j \mathcal{E}_j{}^i + \frac{\gamma}{2} \sum_{i,j} (\mathcal{E}_i{}^j)^2 + \frac{\beta}{2} \mathscr{D}^2 - \frac{(d+1)\beta + \gamma}{2} \mathscr{D} , \quad (2.54)$$

where the operator $\mathscr{L}$ is defined by the expression (2.47) with C^{ij}_{kl} given by eq. (2.45).

2.5   Show that, in the action on functions that depend of W via the vector $\vec{\rho}$, where the stretching exponents are defined by the decomposition (2.30), the operator $\mathscr{L}$ reduces to the operator $\mathcal{L}$ given by eq. (2.50).

2.6   (For volunteers) Prove eq. (2.51) and calculate the constant that appears there. What is the bottom of the spectrum of the operator $\mathcal{H}_{CSM}$?

2.7   Check that the rate function (2.52) satisfies the multiplicative fluctuation relation MFR of Lecture 2.

2.5 Lecture 4.   Generalized flows and dissipative anomaly

Let us consider the simultaneous motion of N Lagrangian particles $\boldsymbol{R}_n(t)$ in the Kraichnan flow. Such particles are correlated due to the fact they move in the same velocity field with long-range spatial correlations. The multi-particle motion will allow new insights into transport properties of the flow. First, we observe that a function $f(\underline{\boldsymbol{R}})$ of the joint positions $\underline{\boldsymbol{R}}(t) \equiv (\boldsymbol{R}_1(t), \ldots, \boldsymbol{R}_N(t))$ evolves according to the Itô SDE

$$\begin{aligned}
\mathrm{d}f(\underline{\boldsymbol{R}}) \;=\; & \sum_{n=1}^{N} v^i(t, R_i) \nabla_{R_n^i} f(\underline{R}) \,\mathrm{d}t \\
& + \tfrac{1}{2} \underbrace{\sum_{m,n=1}^{N} D^{ij}(\boldsymbol{R}_m - \boldsymbol{R}_n) \nabla_{R_m^i} \nabla_{R_n^j} f(\underline{R}) \,\mathrm{d}t}_{\mathcal{M}_N} . \quad (2.55)
\end{aligned}$$

For the expectation values, this gives:

$$\frac{\mathrm{d}}{\mathrm{d}t} \langle f(\underline{\boldsymbol{R}}) \rangle = \langle (\mathcal{M}_N f)(\underline{\boldsymbol{R}}) \rangle . \quad (2.56)$$

In other words, N Lagrangian particles in the Kraichnan flow undergo in mean an effective diffusion with a configuration-dependent diffusivity.

2.5.1 Two-particle dispersion

To start with, let us look at the evolution of a function of the separation vector between two trajectories $\Delta \boldsymbol{R}(t) \equiv \boldsymbol{R}_1(t) - \boldsymbol{R}_2(t)$:

$$\frac{\mathrm{d}}{\mathrm{d}t}\langle f(\Delta\boldsymbol{R})\rangle = \langle (D^{ij}(\boldsymbol{0}) - D^{ij}(\Delta\boldsymbol{R}))\nabla_i\nabla_j f(\Delta\boldsymbol{R})\rangle. \tag{2.57}$$

In the Batchelor regime

$$D^{ij}(\boldsymbol{0}) - D^{ij}(\Delta\boldsymbol{R}) \approx -\frac{1}{2}\Delta R^k \Delta R^l \nabla_k \nabla_l D^{ij}(\boldsymbol{0})$$

and we obtain the same equation as for $\langle f(\delta\boldsymbol{R})\rangle$, see (2.41). In particular, a short calculation gives for the function of the distance $\Delta(t) \equiv |\Delta\boldsymbol{R}(t)|$:

$$\frac{\mathrm{d}}{\mathrm{d}t}\langle f(\Delta)\rangle = \left\langle \left[\frac{2\beta+\gamma}{2}\left(\frac{\mathrm{d}}{\mathrm{d}\ln\Delta}\right)^2 + \lambda_1\frac{\mathrm{d}}{\mathrm{d}\ln\Delta}\right]f(\Delta)\right\rangle, \tag{2.58}$$

with $\lambda_1 = \frac{d-2}{2}\gamma - \beta$ standing for the first Lyapunov exponent, see Problem 4.1 below. For the PDF $P_t(\Delta;\Delta_0)$ of the distance $\Delta(t)$ with $\Delta(0) \equiv \Delta_0$, this gives

$$P_t(\Delta;\Delta_0) = \frac{1}{\sqrt{2\pi(2\beta+\gamma)t}}\exp\left[-\frac{(\ln\frac{\Delta}{\Delta_0} - \lambda_1 t)^2}{2(2\beta+\gamma)t}\right]\frac{1}{\Delta}, \tag{2.59}$$

i.e. a log-normal distribution. It is easy to check that

$$\lim_{\Delta_0\to 0} P_t(\Delta;\Delta_0) = \delta(\Delta), \tag{2.60}$$

see Problem 4.2. This means that when the initial distance Δ_0 of the trajectories tends to zero, so does their time t distance: it takes longer and longer time to separate closer and closer trajectories to the same final distance. Such behavior is compatible with the fact that in each realization of the velocity field, Lagrangian trajectories are uniquely determined by the initial condition. It is also compatible with the exponential growth of the distance of very close trajectories.

Let us now consider the Kraichnan model in the inertial range. Here the difference $D^{ij}(\boldsymbol{0}) - D^{ij}(\boldsymbol{r}) \propto |\boldsymbol{r}|^\xi$ and isotropy fixes its form to

$$D^{ij}(\boldsymbol{0}) - D^{ij}(\boldsymbol{r}) = \beta r^i r^j |\boldsymbol{r}|^{\xi-2} + \frac{1}{2}\gamma\delta^{ij}|\boldsymbol{r}|^\xi \tag{2.61}$$

which, up to normalization, is determined again by the compressibility degree

$$\wp = \lim_{r\to 0}\frac{\nabla_i\nabla_j D^{ij}(\boldsymbol{r})}{\nabla_j\nabla_j D^{ii}(\boldsymbol{r})} = \frac{2(d-1+\xi)\xi^{-1}\beta+\gamma}{2\beta+d\gamma} \tag{2.62}$$

taking values between 0 and 1. The equation for the evolution of the mean value of a function of the time t distance $\Delta(t)$ between two trajectories takes now the form

$$\frac{d}{dt}\langle f(\Delta) \rangle = \left\langle \frac{2\beta + \gamma}{2} \Delta^{\xi - a} \frac{d}{d\Delta} \Delta^a \frac{d}{d\Delta} f(\Delta) \right\rangle \equiv \langle (Mf)(\Delta) \rangle$$

for $a = \frac{(d-1)\gamma}{2\beta + \gamma} = \frac{d+\xi}{1+\wp\xi} - 1$. In other words, $\Delta(t)$ undergoes a one-dimensional diffusion on the half-line $\mathbb{R}_+$ with the generator M and

$$P_t(\Delta; \Delta_0) = e^{tM}(\Delta_0; \Delta). \tag{2.63}$$

The variable $\Delta(t)^{1 - \xi/2} \equiv u(t)$, in turn, undergoes the Bessel diffusion with generator proportional to $u^{1 - D_{\text{eff}}} \partial_u u^{D_{\text{eff}} - 1} \partial_u$, i.e. to the radial Laplacian in D_{eff} dimensions, where

$$D_{\text{eff}} = \frac{2(a + 1 - \xi)}{2 - \xi}$$

takes continuous values. With abuse of language, we could say that $u(t)$ behaves as the norm of the Brownian motion in continuous dimension D_{eff}. As is well known, the short-distance behavior of the Brownian motion below and above two dimensions is different. In particular, for $0 < D_{\text{eff}} < 2$, i.e. for

$$p_c^1 \equiv \frac{d - 2}{2\xi} + \frac{1}{2} < \wp < \frac{d}{\xi^2} \equiv p_c^2,$$

the generator of the Bessel process admits different boundary conditions at $u = 0$ which correspond to different behaviors of Lagrangian particles.

2.5.2 Phases of the Lagrangian flow

There are 3 different phases of trajectory behavior and transport in the inertial range, depending on the level of compressibility [Gaw00, EVa00]. Let us discuss them one by one.

2.5.2.1 Weakly compressible phase

For $0 \leq \wp \leq p_c^1$, i.e. for $D_{\text{eff}} > 2$, there is only one possible boundary condition at zero for M,

$$\left(\Delta^a \frac{d}{d\Delta} f \right)(0) = 0 \tag{2.64}$$

It renders M self-adjoint on $L^2(\mathbb{R}_+, \Delta^{a - \xi} d\Delta)$. Somewhat abusively, we shall call the condition (2.64) reflecting. In reality, zero is an **entrance boundary point** [Bre68] for the diffusion process $\Delta(t)$ if $0 \leq \wp \leq p_c^1$. It may never be attained from positive values. The latter may, however, be reached by starting from zero. Here

$$\lim_{\Delta_0 \to 0} P_t(\Delta; \Delta_0) = \text{const.}\, t^{-\frac{a+1-\xi}{2-\xi}} \exp\left[-\frac{2\Delta^{2-\xi}}{(2\beta+\gamma)(2-\xi)^2 t}\right] \qquad (2.65)$$

signifying that the time t distance of two Lagrangian particles stays smoothly distributed in the limit when the initial distance approaches zero. Consequently, particle trajectories separate from arbitrary close points in finite time (unlike in the chaotic regime with exponential separation!). This implies a rather exotic property of the Lagrangian flow in the limit when the inertial range extends to arbitrary small distances (modeling the infinite Reynolds number velocities): its **spontaneous randomness**. The deterministic PDFs $\delta(r - R(t; t_0, r_0 | v))$ of the time t trajectory positions in fixed velocity realizations v are replaced for infinite Re by the PDFs $P(t, r; t_0, r_0 | v)$ that are not concentrated at single points. This is possible since, for $(D(0) - D(r)) \propto |r|^\xi$, typical velocities $v(t, r)$ are only Hölder continuous in space with exponents smaller than $\xi/2 < 1$ and the solutions of the trajectory equation (2.37) *are not* uniquely determined at fixed v by the initial positions r_0, see Fig. 2.2†. Instead, they form a **generalized flow**: a stochastic (Markov) process with transition probabilities $P(t, r; t_0, r_0 | v)$ that have been constructed by Le Jan and Raimond in [LeJ02].

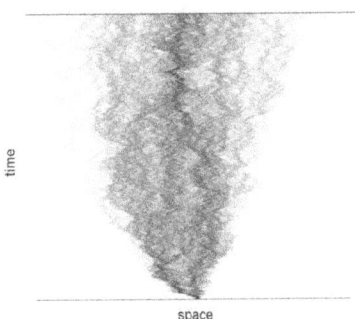

Fig. 2.2. Behavior of Lagrangian trajectories in a rough weakly compressible velocity

2.5.2.2 Strongly compressible phase

For $\wp > p_c^2$. i.e. for $D_{\text{eff}} < 0$, the only possible boundary condition for M is

$$f(0) = 0. \tag{2.66}$$

It renders the operator M self-adjoint in the space $L^2(\mathbb{R}_+, \Delta^{a-\xi}d\Delta)$. We shall call the condition (2.66) absorbing. Here, zero is an **exit boundary point** [Bre68] for the process $\Delta(t)$ which can attain it but cannot come back from it. For $\Delta_0 > 0$,

$$P_t(\Delta; \Delta_0) = \text{regular} + \text{const.}\, \delta(\Delta) \tag{2.67}$$

In this phase, there is a positive probability for the Lagrangian trajectories that start at a distance $\Delta_0 > 0$ to collapse together by time t, see Fig. 2.3†. The trapping of Lagrangian particles, that increases with increase of the compressibility degree $\wp$, finds in the strongly compressible phase quite a dramatic manifestation, as indicated by the presence of the contact term on the right hand side of eq. (2.67). When $\Delta_0 \to 0$, then the contact term becomes dominant:

$$\lim_{\Delta_0 \to 0} P_t(\Delta; \Delta_0) = \delta(\Delta).$$

The latter behavior reflects the deterministic character of Lagrangian trajectories in the strongly compressible phase:

$$P(t, \boldsymbol{r}; t_0, \boldsymbol{r}_0 | \boldsymbol{v}) = \delta(\boldsymbol{r} - \boldsymbol{R}(t; t_0, \boldsymbol{r}_0 | \boldsymbol{v}))$$

for trajectories $\boldsymbol{R}(t; t_0, \boldsymbol{r}_0 | \boldsymbol{v})$ uniquely determined by the initial condition but, nevertheless, collapsing together at later times (again an exotic behavior).

2.5.2.3 Intermediate phase

Finally, for $p_c^1 < \wp < p_c^2$, i.e. for $0 < D_{\text{eff}} < 2$, zero is a **regular boundary point** [Bre68] for the process $\Delta(t)$. Both reflecting and absorbing boundary conditions for M are possible, leading to generalized or deterministic Lagrangian flows, respectively. The reflecting condition is selected by adding very small diffusivity, the absorbing one by considering velocities smeared at very small distances (mimicking the effect of small viscosity) [EVa00]. The other permitted boundary conditions are the "sticky" ones [Gaw04]:

$$\mu\left(\Delta^{\xi-a}\frac{\mathrm{d}}{\mathrm{d}\Delta}\Delta^a\frac{\mathrm{d}}{\mathrm{d}\Delta}f\right)(0) = \left(\Delta^a\frac{\mathrm{d}}{\mathrm{d}\Delta}f\right)(0) \tag{2.68}$$

† Reprinted with permission from [Fal01][http://link.aps.org/abstract/RMP/v73/p913]. Copyright 2000 by the American Physical Society.

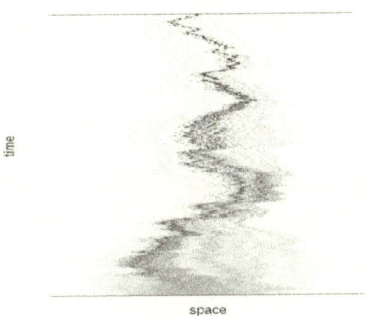

Fig. 2.3. Behaviour of Lagrangian trajectories in a rough strongly compressible velocity field.

for μ being the measure of stickiness, $0 < \mu < \infty$. These conditions render M self-adjoint on $L^2(\mathbb{R}_+, (\Delta^{a-\xi} + \mu\delta(\Delta))\mathrm{d}\Delta)$. They lead to generalized flows in which particles stick to each other and separate again, spending together a finite portion of the total time but never a finite time-interval. The sticky condition corresponds to fine-tuned small diffusivity and small viscosity [EVa01, Gaw04].

For $p_c^1 < \wp < p_c^2$, the transition probabilities $P(t, \boldsymbol{r}; t_0, \boldsymbol{r}_0 | \boldsymbol{v})$ have been constructed in [LeJ04] both for the reflecting boundary condition and for the absorbing one. In the first case, they depend on the velocities $\boldsymbol{v}$ (that are white-noise in time), whereas in the second case, on the (white noise) $\boldsymbol{v}$ and on an additional (non-standard) **black noise** [Tsi04] that decides what the particles do when they meet (which trajectory they follow together).

2.5.3 Scalar cascades

The exotic behaviors of Lagrangian trajectories in Hölder continuous Kraichnan velocities are source of important transport phenomena. We shall concentrate below on the scalar transport, see also Sect. 1.6.1 of the course of G. Falkovich. One of the implications of the advection-diffusion equation (2.6) in incompressible velocities is the balance of scalar "energy" $\int \theta^2$,

$$\frac{\mathrm{d}}{\mathrm{d}t} \int \theta(t, \boldsymbol{r})^2 \, \mathrm{d}\boldsymbol{r} = -2\kappa \int (\boldsymbol{\nabla}\theta(t, \boldsymbol{r}))^2 \, \mathrm{d}\boldsymbol{r} + 2 \int \theta(t, \boldsymbol{r}) \, g(t, \boldsymbol{r}) \, \mathrm{d}\boldsymbol{r} \,,$$

with the first term on the right hand side describing the dissipation and the second one the injection of scalar energy by the source. One could naively expect that the scalar energy is conserved for $g = 0$ in the limit $\kappa \to 0$. In this limit, the evolution of the decaying (i.e. sourceless) scalar is given by the equation

$$\theta(t, \boldsymbol{r}) = \int P(t_0, \boldsymbol{r}_0; t, \boldsymbol{r} | \boldsymbol{v})\, \theta(t_0, \boldsymbol{r}_0)\, \mathrm{d}\boldsymbol{r}_0\,, \tag{2.69}$$

as indicated by eq. (2.9) (we shall look more closely at the evolution of scalar passively advected by Kraichnan velocities in Lecture 5). In an incompressible flow, one has

$$\int P(t_0, \boldsymbol{r}_0; t, \boldsymbol{r} | \boldsymbol{v})\, \mathrm{d}\boldsymbol{r}_0 = \int P(t_0, \boldsymbol{r}_0; t, \boldsymbol{r} | \boldsymbol{v})\, \mathrm{d}\boldsymbol{r} = 1\,.$$

It follows, using eq. (2.69) in passing from the first to the second line below to develop the mixed term in the expansion of the square, that

$$
\begin{aligned}
0 \leq \int \mathrm{d}\boldsymbol{r} \int P(t_0, \boldsymbol{r}_0; t, \boldsymbol{r} | \boldsymbol{v}) \left(\theta(t_0, \boldsymbol{r}_0) - \theta(t, \boldsymbol{r}) \right)^2 \mathrm{d}\boldsymbol{r}_0 \\
= \int \theta(t_0, \boldsymbol{r}_0)^2\, \mathrm{d}\boldsymbol{r}_0 - \int \theta(t, \boldsymbol{r})^2\, \mathrm{d}\boldsymbol{r}
\end{aligned}
$$

so that the total scalar energy cannot grow. Moreover, it is conserved if and only if, for each $\boldsymbol{r}$, $\theta(t_0, \boldsymbol{r}_0) = \theta(t, \boldsymbol{r})$ on the support of $P(t_0, \cdot\,; t, \boldsymbol{r} | \boldsymbol{v})$, i.e. if and only if the flow is deterministic in fixed velocity fields. Consequently, in a generalized incompressible flow, there is **persistent dissipation** of the scalar energy.

A similar effect may be seen also in the presence of weak compressibility in averaged quantities. Suppose that, at time zero, we are given a homogeneous isotropic scalar distribution with the 2-point function

$$\langle \theta(0, \boldsymbol{r}_0)\, \theta(0, \boldsymbol{0}) \rangle \equiv F_0^{(2)}(\Delta_0)\,.$$

where $\Delta_0 = |\boldsymbol{r}_0|$. Recall from eq. (2.9) that the scalar evolution is determined by the backward Lagrangian flow. If the initial scalar distribution is independent of velocities then the decaying scalar 2-point function at the later time t is given by

$$F_t^{(2)}(\Delta) = \int P_{-t}(\Delta_0; \Delta)\, F_0^{(2)}(\Delta_0)\, \mathrm{d}\Delta_0\,.$$

In particular, for the mean scalar energy density $\langle \theta(t, \boldsymbol{0})^2 \rangle = F_t^{(2)}(0)$, we obtain

$$F_t^{(2)}(0) = \int P_{-t}(\Delta_0; 0)\, F_0^{(2)}(\Delta_0)\, \mathrm{d}\Delta_0\,.$$

Since $F_0^{(2)}(\Delta_0) \leq F_0^{(2)}(0)$ by the Schwarz inequality, it follows that the mean scalar energy density is non-increasing. It is conserved only if $P_{-t}(\Delta_0; 0) = \delta(\Delta_0)$, i.e. when the Lagrangian trajectories are uniquely determined by their final position.

If the scalar is forced with an independent random Gaussian source g with mean zero and covariance

$$\langle g(t, \mathbf{r}) \, g(t', \mathbf{r}') \rangle = \delta(t - t') \, \chi(|\mathbf{r} - \mathbf{r}'|) \tag{2.70}$$

for $0 \leq \chi(\Delta) \leq \chi(0)$ then the scalar 2-point function evolves according to the formula

$$\begin{aligned}
F_t^{(2)}(\Delta) &= \int P_{-t}(\Delta_0; \Delta) \, F_0^{(2)}(\Delta_0) \, \mathrm{d}\Delta_0 \\
&+ \int_0^t \mathrm{d}s \int P_{-t+s}(\Delta_0; \Delta) \, \chi(\Delta_0) \, \mathrm{d}\Delta_0 \, .
\end{aligned}$$

In the inertial range of the Kraichnan model, due to the time reversibility of the velocity ensemble, $P_{-t}(\Delta_0; \Delta) = P_t(\Delta_0; \Delta) = e^{tM}(\Delta; \Delta_0)$, see eq. (2.63). Consequently, the forced scalar 2-point function solves the differential equation

$$\frac{\mathrm{d}}{\mathrm{d}t} F_t^{(2)}(\Delta) = (M F_t^{(2)})(\Delta) + \chi(\Delta) \, . \tag{2.71}$$

Upon setting $\Delta = 0$ in the last relation, we obtain the energy balance with $-(M F_t^{(2)})(0)$ representing the mean scalar energy dissipation rate and $\chi(0)$ the mean scalar energy injection rate. Indeed, one can show that

$$-(M F_t^{(2)})(0) = \lim_{\kappa \to 0} 2\kappa \langle (\nabla \theta(t, \mathbf{0}))^2 \rangle \, , \tag{2.72}$$

where $2\kappa \langle (\nabla \theta(t, \mathbf{0}))^2 \rangle$ is the mean dissipation rate in the presence of diffusion. In the weakly compressible phase, forced advected scalar reaches a stationary state in which the dissipation rate balances the injection rate. In particular, the former is non-vanishing in spite of being the $\kappa \to 0$ limit of a quantity that involves an explicit factor κ. This is the so called **dissipative anomaly** which is due to the spontaneous randomness of the Lagrangian flow. It conditions the appearance of a **direct cascade** of scalar energy towards short-distance scales in the weakly compressible phase, compare to Sect. 1.4.1 of the course of G. Falkovich. In this phase, the stationary scalar structure function

$$\langle (\theta(t, \mathbf{r}) - \theta(t, \mathbf{0})^2 \rangle = 2(F_t^{(2)}(0) - F_t^{(2)}(\Delta)) \equiv S^{(2)}(\Delta)$$

is proportional to $\Delta^{2-\xi}$ for small Δ exhibiting the normal scaling predicted by the dimensional analysis, see Problem 4.3 below.

In the strongly compressible phase, where the Lagrangian flow is deterministic, there is no persistent dissipation. In the decaying case, the mean scalar energy density $F_t^{(2)}(0)$ is conserved whereas in the forced case, it grows linearly in time with the rate equal to $\chi(0)$. Now, it is the term $\frac{d}{dt}F_t^{(2)}(0)$ that balances the mean injection rate in eq. (2.71), whereas the mean dissipation rate $-(MF_t^{(2)})(0)$ vanishes. The transported scalar exhibits an **inverse cascade** of scalar energy towards long-distance scales. The evolution equation (2.71) induces the one for the 2-point structure function

$$\frac{d}{dt}S_t^{(2)}(\Delta) = (MS_t^{(2)})(\Delta) + 2(\chi(0) - \chi(\Delta)). \qquad (2.73)$$

Although $F_t^{(2)}(\Delta)$ does not have a long time limit, $S_t^{(2)}(\Delta)$ reaches a stationary form with an anomalous scaling $\propto \Delta^{1-a}$ for small Δ, see Problem 4.3 below.

In the intermediate phase, and in the presence of forcing, one obtains a direct scalar energy cascade for the reflecting boundary condition that leads to a generalized Lagrangian flow, and the inverse cascade for the absorbing boundary condition for which the Lagrangian flow is deterministic. The mean energy density $F_t^{(2)}(0)$ grows indefinitely in both cases but the scalar structure function reaches a stationary form proportional to $\Delta^{2-\xi}$ or to Δ^{1-a}, respectively. For the sticky conditions, which also lead to the spontaneous randomness of the Lagrangian flow, the persistent dissipation is still present, causing a direct energy cascade of the forced scalar. For such conditions, the 2-point scalar structure function reaches a stationary form with an anomalous scaling $\propto \Delta^{1-a}$ for small Δ.

Conclusion. *In the inertial range, the Kraichnan model Lagrangian particles exhibit exotic behaviors that are impossible in differentiable dynamical systems but should be common in non-differentiable dynamical systems $\frac{dX}{dt} = \mathcal{X}(t, X)$ with $\mathcal{X}(t, \cdot)$ only Hölder continuous. These behaviors include explosive separation of trajectories related to spontaneous generation of randomness and to persistent dissipation in passive advection, trajectory stickiness, or implosive collapse of deterministic trajectories. Such unconventional behaviors of trajectories lead to non-equilibrium transport phenomena like the direct or inverse cascades with fluxes of conserved quantities.*

2.5.4 Problems

2.1   Prove the relations (2.58) and (2.59).

2.2   Establish the convergence (2.60).

2.3   For the forced scalar with the 2-point correlation function satisfying eq. (2.71), find explicit expressions for the stationary 2-point structure function in the weakly compressible and in the strongly compressible phase. What is their short-distance scaling?

2.6 Lecture 5.   Zero-mode scenario for intermittency

We shall look more closely at the evolution of the scalar field passively advected by the Kraichnan velocities. In particular, we shall exploit the relations between the motion of N Lagrangian particles and N-point scalar correlation functions in the inertial range of scales. This will permit to address the problem of intermittency, an important phenomenon in turbulence characterized by frequent appearance of strong signals interwoven with weak ones both in time and in space domain. The discussion will treat in a more formal way the ideas sketched in Sect. 1.5 of the course of G. Falkovich.

2.6.1 Stochastic PDE for scalar

First, suppose that the velocities are smooth in space and the scalar stays constant along Lagrangian trajectories that solve the Itô SDE (2.37) so that

$$\theta(t, \boldsymbol{R}(t; t_0, \boldsymbol{r}_0)) = \theta(t_0, \boldsymbol{r}_0) = \text{const.}$$

In the Kraichnan velocity ensemble this means that the scalar field has to satisfy the Itô stochastic PDE

$$\mathrm{d}_t \theta(t, \boldsymbol{r}) = -\boldsymbol{v}(t, \boldsymbol{r}) \, \mathrm{d}t \cdot \boldsymbol{\nabla} \theta(t, \boldsymbol{r}) + \tfrac{1}{2} D_0 \boldsymbol{\nabla}^2 \theta(t, \boldsymbol{r}) \, \mathrm{d}t \qquad (2.74)$$

Indeed, the Itô rule gives then

$$\begin{aligned}
\mathrm{d}_t \theta(t, \boldsymbol{R}(t)) =& \big(\theta(t + \mathrm{d}t, \boldsymbol{R}(t + \mathrm{d}t)) - \theta(t, \boldsymbol{R}(t + \mathrm{d}t))\big) \\
& + \big(\theta(t, \boldsymbol{R}(t + \mathrm{d}t)) - \theta(t, \boldsymbol{R}(t))\big) \\
=& - \boldsymbol{v}(t, \boldsymbol{R}(t)) \, \mathrm{d}t \cdot \boldsymbol{\nabla} \theta(t, \boldsymbol{R}(t)) + \tfrac{1}{2} D_0 \boldsymbol{\nabla}^2 \theta(t, \boldsymbol{R}(t)) \, \mathrm{d}t \\
& + \boldsymbol{v}(t, \boldsymbol{R}(t)) \, \mathrm{d}t \cdot \boldsymbol{\nabla} \theta(t, \boldsymbol{R}(t)) + \tfrac{1}{2} D_0 \boldsymbol{\nabla}^2 \theta(t, \boldsymbol{R}(t)) \, \mathrm{d}t \\
& - D_0 \boldsymbol{\nabla}^2 \theta(t, \boldsymbol{R}(t)) \, \mathrm{d}t = 0
\end{aligned}$$

In the intermediate expression, the contribution on the third line comes from the time increment (2.74) of the field θ, on the fourth line, from the time differential of $\boldsymbol{R}(t)$ and, on the fifth line, from the combination of the increments of $\theta(t)$ and of $\boldsymbol{R}(t)$ proportional to $\boldsymbol{v}$. The Stratonovich stochastic PDE corresponding to the Itô one (2.74) would be

$$d_t\theta(t,\boldsymbol{r}) = -\boldsymbol{v}(t,\boldsymbol{r})\,dt \circ \cdot \boldsymbol{\nabla}\theta(t,\boldsymbol{r})$$

(i.e. without the diffusive term). For the forced scalar, one obtains the Itô equation

$$d_t\theta(t,\boldsymbol{r}) = -\boldsymbol{v}(t,\boldsymbol{r})\,dt \cdot \boldsymbol{\nabla}\theta(t,\boldsymbol{r}) + \tfrac{1}{2}D_0\boldsymbol{\nabla}^2\theta(t,\boldsymbol{r})\,dt + g(t,\boldsymbol{r})\,dt$$

and the corresponding Stratonovich one without the diffusive term.

2.6.2 Evolution of scalar correlation functions

Suppose now that the source $g(t,\boldsymbol{r})$ is an independent white in time noise with the covariance (2.70). The last equation induces then for the product $\prod_{n=1}^{N}\theta(t,\boldsymbol{r}_n)$ the Itô equation

$$
\begin{aligned}
d_t\prod_{n=1}^{N}\theta(t,\boldsymbol{r}_n) = {}&-\sum_{n=1}^{N}\boldsymbol{v}(t,\boldsymbol{r}_n)\,dt\cdot\boldsymbol{\nabla}\theta(t,\boldsymbol{r}_n)\prod_{n'\neq n}\theta(t,\boldsymbol{r}_{n'}) \\
&+\sum_{n=1}^{N}\tfrac{1}{2}D_0\boldsymbol{\nabla}^2\theta(t,\boldsymbol{r}_n)\,dt\prod_{n'\neq n}\theta(t,\boldsymbol{r}_{n'}) \\
&+\sum_{m<n}D^{ij}(\boldsymbol{r}_m-\boldsymbol{r}_n)\nabla_i\theta(t,\boldsymbol{r}_m)\nabla_j\theta(t,\boldsymbol{r}_n)\,dt\prod_{n'\neq m,n}\theta(t,\boldsymbol{r}_{n'}) \\
&+\sum_{n=1}^{N}g(t,\boldsymbol{r}_n)\,dt\prod_{n'\neq n}\theta(t,\boldsymbol{r}_{n'}) \\
&+\sum_{m<n}\chi(|\boldsymbol{r}_m-\boldsymbol{r}_n|)\,dt\prod_{n'\neq m,n}\theta(t,\boldsymbol{r}_{n'}),
\end{aligned}
$$

or for the expectation value $\left\langle \prod_{n=1}^{N}\theta(t,\boldsymbol{r}_n)\right\rangle \equiv F_t^{(N)}(\underline{\boldsymbol{r}})$,

$$
\begin{aligned}
\frac{d}{dt}F_t^{(N)}(\underline{\boldsymbol{r}}) = {}&\mathcal{M}_N F_t^{(N)}(\underline{\boldsymbol{r}}) \\
&+\sum_{m<n}\chi(|\boldsymbol{r}_m-\boldsymbol{r}_n|)\,F_t^{(N-2)}(\boldsymbol{r}_1,..,\hat{\boldsymbol{r}_m},..,\hat{\boldsymbol{r}_n},..,\boldsymbol{r}_N),\quad (2.75)
\end{aligned}
$$

where the over-hat signals that the term should be omitted, compare to Eq. (1.27) of the course of G. Falkovich. For $N=2$ and in the inertial range, this is eq. (2.71). The operators $\mathcal{M}_N$ are the same as in (2.55). At least in principle, the evolution equations (2.75) allow to find the (equal time) scalar correlation functions $F_t^{(N)}(\boldsymbol{r})$ inductively. This is another simplifying feature of the Kraichnan model (usually, the time derivative of the N-point function cannot be expressed only by the lower-point functions). If the scalar reaches at long times a stationary state, then in the latter,

$$\mathcal{M}_N F_t^{(N)}(\boldsymbol{r}) = -\sum_{m<n} \chi(|\boldsymbol{r}_m - \boldsymbol{r}_n|) \, F_t^{(N-2)}(\boldsymbol{r}_1, .., \hat{\boldsymbol{r}}_m, .., \hat{\boldsymbol{r}}_n, .., \boldsymbol{r}_N) \,. \quad (2.76)$$

In the action on translation-invariant functions $f(\boldsymbol{r}_1, \ldots, \boldsymbol{r}_N)$ satisfying $\sum_{n=1}^{N} \nabla_{r_n^i} f(\boldsymbol{r}) = 0$ for all i,

$$\mathcal{M}_N f(\boldsymbol{r}) = \tfrac{1}{2} \sum_{m\neq n} \left(D^{ij}(\boldsymbol{r}_m - \boldsymbol{r}_n) - D^{ij}(\boldsymbol{0}) \right) \nabla_{r_m^i} \nabla_{r_n^j} f(\boldsymbol{r}) \,, \quad (2.77)$$

see eq. (2.55). Hence in the inertial range, where $(D^{ij}(\boldsymbol{0}) - D^{ij}(\boldsymbol{r})) \propto |\boldsymbol{r}|^\xi$, the operator $\mathcal{M}_N$ scales as $(distance)^{\xi-2}$. Since $\chi(\boldsymbol{r}) = O(1)$ for small $|\boldsymbol{r}|$, the short-range scaling of the correlation functions $F_t^{(N)}(\boldsymbol{r})$ as $(distance)^{\frac{N}{2}(2-\xi)}$ would be consistent with eq. (2.76). This is the prediction of the naive dimensional analysis.

2.6.3 Zero modes

Note that one could add to $F_t^{(N)}$ any zero mode $f^{(N)}$ of $\mathcal{M}_N$ satisfying

$$\mathcal{M}_N f^{(N)} = 0$$

and still have eqs. (2.76) fulfilled. Such zero modes may indeed contribute to the correlation functions and spoil the inertial range scaling behavior predicted by the dimensional analysis. The simplest example is provided by the 2-point function in the weakly compressible phase since $\lim_{\Delta \to 0} F^{(2)}(0) > 0$ and it is only the 2-point structure function $S^{(2)}(\Delta) = 2(F^{(2)}(\Delta) - F^{(2)}(0))$ that scales in agreement with the dimensional prediction as $\Delta^{2-\xi}$, see Problem 4.3 above. Hence the 2-point correlation function exhibits the dimensional scaling only after subtraction from it of a trivial (constant) zero mode of $\mathcal{M}_2$.

One could expect that similar situation persists for N-point scalar correlation function and the dimensional scaling holds for the N-point structure functions

$$S^{(N)}(\Delta) = \left\langle (\theta(t,\boldsymbol{r}) - \theta(t,\boldsymbol{0}))^N \right\rangle \qquad \text{with} \qquad \Delta = |\boldsymbol{r}|.$$

It appears, however, that this is not the case [Che95, Gaw95, Shr95]. In the stationary state with direct scalar energy cascade reached at weak compressibility, the N-point scalar structure functions with even $N > 2$ exhibit anomalous inertial range scaling

$$S^{(N)}(\Delta) \propto \Delta^{\zeta_N} \tag{2.78}$$

for small Δ with $\zeta_N < \frac{N}{2}(2 - \xi)$. The scaling exponent is determined by the contribution to $F^{(N)}(\underline{r})$ coming from scaling zero-modes $f^{(N)}$ of $\mathcal{M}_N$ such that

$$f^{(N)}(\lambda\underline{r}) = \lambda^{\zeta_N} f^{(N)}(\underline{r}).$$

In order to interpret this zero-mode scenario of anomalous scaling, recall from Lecture 4 that the mean values of functions of N Lagrangian trajectories evolve according eq. (2.56). Zero-modes of $\mathcal{M}_N$ correspond to functions f such that $f(\underline{R}(t))$ are **martingales** of the N-particle process, i.e. are conserved in mean [Ber98]:

$$\left\langle f(\underline{R}(t)) \right\rangle_{\underline{R}(0) = \underline{R}_0} = f(\underline{R}_0),$$

In general, such many-particle **statistical conservation laws**, that break the scaling symmetry in the Lagrangian evolution, describe subtle memory effects, see the lectures of G. Falkovich in this volume. They are usually hidden and, to expose them, one has to resort to perturbative or numerical analysis. For generic scaling functions $f(\lambda\underline{R}) = \lambda^\zeta f(\underline{R})$ that are translation-invariant, the average $\left\langle f(\underline{R}(t)) \right\rangle$ grows in time as $t^{\frac{\zeta}{2-\xi}}$. For example,

$$\left\langle |\boldsymbol{r}_m(t) - \boldsymbol{r}_n(t)|^2 \right\rangle \propto t^{\frac{2}{2-\xi}}$$

at long times, as may be inferred from eq. (2.65). For the Komogorov value $\xi = \frac{4}{3}$, this gives the Richardson behavior $\propto t^3$ observed in turbulent flows [Ric26, Bif05]. Conservation of the average of $f(\underline{R}(t))$ requires non-generic functions that exist only for certain scaling dimensions. Besides, the zero modes $f^{(N)}$ that dominate the structure functions have to be irreducible in the sense that they are not combinations of functions that depend on smaller number of variables because such combinations do not contribute to the structure functions. See the lectures of G. Falkovich in this volume for the geometric interpretation of such zero modes.

The scaling zero-modes of $\mathcal{M}_N$ that provide the dominant contributions to the structure function $S^{(N)}(\Delta)$ were found in the leading order of the perturbation expansion in powers of ξ [Gaw95, Ber96] and in powers of $\frac{1}{d}$

[Che95, Che96] during the years 1995-96. The result gives for the scaling exponents with even N:

$$
\zeta_N = \begin{cases} \underbrace{\frac{N}{2}(2-\xi)}_{} - \underbrace{\frac{N(N-2)(1+2\wp)\,\xi}{2(d+2)} + O(\xi^2)}_{} \\ \underbrace{\frac{N}{2}(2-\xi)}_{} - \underbrace{\frac{N(N-2)(1+2\wp)\,\xi}{2d} + O(\frac{1}{d^2})}_{} \end{cases} \tag{2.79}
$$

dimensional   anomaly
scaling

The perturbative result has been confirmed by numerical simulations [Fri99], see Fig. 2.4† By now, terms to the third order are known thanks to the work of the St. Petersburg group [Adz01].

The fact that the exponents ζ_N are smaller than the dimensional prediction $\frac{N}{2}(\xi-2) = \frac{N}{2}\zeta_2$ for $N > 2$ means that

$$
\frac{S^{(N)}(\Delta)}{S^{(2)}(\Delta)^{N/2}} \quad \propto \quad \Delta^{-\left(\frac{N}{2}\zeta_2-\zeta_N\right)}
$$

becomes large at small distance. This signals the presence of long tails in the stationary distribution of the scalar differences $\theta(t,\boldsymbol{r}) - \theta(t,\boldsymbol{0})$ that get more and more pronounced for small $\Delta = |\boldsymbol{r}|$. Such tails indicate that large values of the scalar difference occur relatively frequently and signal small scale **intermittency** of the advected scalar. The anomalous scaling had been indeed observed in the structure functions of scalars advected by realistic flows [Ant84, Moi01], but also in the turbulent velocity structure functions, see the lectures of G. Falkovich in this volume. Zero-modes were shown numerically to be responsible for intermittency of the passive scalar transport in the inverse cascade of the two-dimensional Navier-Stokes turbulence [Cel01a]. It is plausible that a similar scenario accounts for observed scalar intermittency in real three-dimensional turbulent flows.

Conclusion. *The Kraichnan model allowed to associate intermittency of the transported quantities with the presence of hidden statistical conservation laws, the "zero modes". The zero-mode scenario realized in this model permitted perturbative calculation of the anomalous scaling exponents of the scalar structure functions. There are indications that similar mechanism is responsible for intermittency is realistic situations.*

---

† Reprinted with permission from [Fri99]. Copyright 1999, American Institute of Physics.

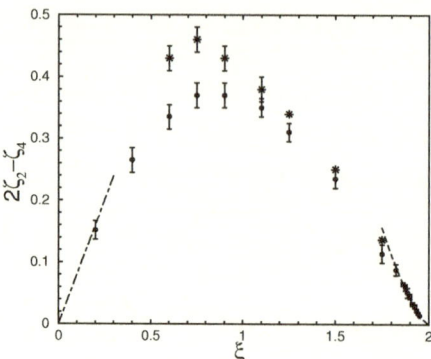

Fig. 2.4. Numerical results for $2\zeta_2 - \zeta_4$ as a function of ξ in two- (upper points) and three-dimensional (lower points) incompressible Kraichnan model, with the broken line in the left corner representing the perturbative result (2.79) for $d = 3$, from [Fri99]

2.7 Problems

2.1   Show that, for $(D^{ij}(\mathbf{0}) - D^{ij}(\mathbf{r}))$ given by eq. (2.61), the operator $\mathcal{M}_N$ in the action on translation-invariant functions $f(\underline{\mathbf{r}})$ of N particle positions becomes proportional to the (Nd)-dimensional Laplacian in the limit $\xi \to 0$ at constant $\wp$.

2.2   Check that

$$f_0^4(\mathbf{r}_1, .., \mathbf{r}_4) = 2(d+2)|\mathbf{r}_1 - \mathbf{r}_2|^2|\mathbf{r}_3 - \mathbf{r}_4|^2 - d(|\mathbf{r}_1 - \mathbf{r}_2|^4 + |\mathbf{r}_3 - \mathbf{r}_4|^4)$$

is a zero mode in this limit. In fact, it is the $\xi \to 0$ limit of the scaling zero mode f^4 that gives the dominant short-distance contribution to the 4-point structure function $S^4(\mathbf{r})$ [Ber96].

2.3   Suppose that, in the stationary state, the PDF of the scalar field difference were of the log-normal form satisfying

$$P_\Delta(\vartheta) \equiv \left\langle \delta(\vartheta - \theta(t, \mathbf{r}) + \theta(t, \mathbf{0})) \right\rangle = e^{-\frac{(\ln|\vartheta| - B)^2}{2A}} \frac{1}{\sqrt{8\pi A}\,\vartheta} \quad (2.80)$$

for $|\mathbf{r}| \equiv \Delta < \Delta_0$ with

$$A = -\frac{(1+2\wp)\,\xi}{d+2} \ln \frac{\Delta}{\Delta_0}, \qquad B = \left(\frac{2-\xi}{2} + \frac{(1+2\wp)\,\xi}{d+2}\right) \ln \frac{\Delta}{\Delta_0}.$$

Show that the structure functions would then exhibit the scaling behavior (2.78) for $\Delta < \Delta_0$ with anomalous exponents consistent with the perturbative result (2.79).

2.8 End remarks

Let us briefly summarize the main points of these lectures.

Passive transport in turbulent velocities is related to the flow of Lagrangian or inertial particles carried by the fluid. At the scales where velocities are smooth in space (intermediate Reynolds numbers or Batchelor regime), such flows may be viewed as examples of random differentiable dynamical systems and may be studies with the standard tools of the theory of chaotic dynamical systems that relate to the local stretching and contraction around trajectories. A quantitative description of the transport phenomena requires, nevertheless, a subtle information about rare events (multiplicative large deviations).

In the regime of high Reynolds numbers (or in the inertial range), the motion of particles provides, instead, examples of random non-differentiable dynamical systems. As the work on the Kraichnan model has shown, such systems lead to unconventional flows with explosive trajectory separation or implosive trajectory trapping. These exotic behaviors have important effect on the transport properties and condition appearance of direct or inverse cascades of conserved quantities. The intermittency of passively advected fields in the direct cascade, signaled by the breakdown of the scale invariance, may be traced back to the presence of hidden statistical conservation laws of the particle motion.

It is expected that at least some of the exotic behaviors of the Lagrangian flow found in the inertial range Kraichnan model persist in flows with temporal correlations [Sok99, Cha03, Bif05], but the issue requires certainly further study. An extension of the analysis of the dynamics of Lagrangian particles in the Kraichnan model to the case of particles with inertia is a subject of active research [Meh05, Hor05, Bec06]. Inertial particles in realistic flows are also intensively studied in numerical simulations, see e.g. [Bec07]. A fundamental problem for applications is the effect on the transport phenomena of non-linearities in the advection-diffusion equations accounting for chemical or biological reactions, see [Tel05] for a review and [Che03] for a case reducible to linear equation. If the diffusion is sufficiently strong, such questions could be studied by extending the renormalization group methods discussed in J. Cardy's lectures in this volume to the advection-reaction-diffusion equations. Note in passing that the intermittency in the passive scalar advection in the Kraichnan model with rough velocities have been analyzed within the field-theoretic renormalization group approach, see [Adz02] and references therein.

The zero mode scenario is believed to be relevant for describing intermittency in the statistics of quantities obeying linear evolution equations, even if the latter are not directly related to the Lagrangian flow [Ara01]. Whether similar scenarios may describe intermittency of turbulent velocities themselves remains to be seen. There are some encouraging signs that it might be the case [Ang06, Ben06].

2.9 Solutions of problems

2.9.1 Problems to Lecture 1

Problem 1.1   The SDE (2.17) for the continuous function $\boldsymbol{R}(t)$ is equivalent to the integral equation

$$\boldsymbol{R}(t) - \boldsymbol{R}(t_0) = \int_{t_0}^{t} \boldsymbol{v}(s, \boldsymbol{R}(s))\, \mathrm{d}s + \sqrt{2\kappa}\,(\boldsymbol{\beta}(t) - \boldsymbol{\beta}(t_0))\,. \qquad (2.81)$$

The latter does not require any convention, but it should be born in mind that $\boldsymbol{\beta}(t) - \boldsymbol{\beta}(t_0)$ is of the order $|t - t_0|^{\frac{1}{2}}$ for small $|t - t_0|$ because it has vanishing mean and

$$\overline{[\beta^i(t) - \beta^i(t_0)][\beta^j(t) - \beta^j(t_0)]} = \delta^{ij}\,|t - t_0|\,.$$

In contrast, the first term on the right hand side of eq. (2.81) is of the order of $|t - t_0|$ and is equal to

$$\boldsymbol{v}(t_0, \boldsymbol{R}(t_0))(t - t_0) + o(|t - t_0|)$$

with $o(|t - t_0|)/|t - t_0|$ going to zero when $t \to t_0$. It follows that

$$\boldsymbol{R}(t + \Delta t) - \boldsymbol{R}(t) = \boldsymbol{v}(t, \boldsymbol{R}(t))\Delta t + \sqrt{2\kappa}(\boldsymbol{\beta}(t + \Delta t) - \boldsymbol{\beta}(t)) + o(\Delta t)\,.$$

For $t_0 < t_1 < \cdots < t_n < t_{n+1} \equiv t$ with $t_{m+1} - t_m = \Delta t$, we may write the increment of f as a telescoping sum and expand to the second order to obtain the relations

$$f(\boldsymbol{R}(t)) - f(\boldsymbol{R}(t_0)) = \sum_{m=0}^{n} [f(\boldsymbol{R}(t_{m+1})) - f(\boldsymbol{R}(t_m))]$$

$$= \sum_{m=0}^{n} [\boldsymbol{R}(t_{m+1}) - \boldsymbol{R}(t_m)] \cdot \boldsymbol{\nabla} f(\boldsymbol{R}(t_m))$$

$$+ \kappa \sum_{m=0}^{n} [\beta^i(t_{m+1}) - \beta^i(t_m)][\beta^j(t_{m+1}) - \beta^j(t_m)]\nabla_i \nabla_j f(\boldsymbol{R}(t_m))$$

$$+ o(\Delta t)$$

$$= \sum_{m=0}^{n} \boldsymbol{v}(t_m, \boldsymbol{R}(t_m)) \cdot (\boldsymbol{\nabla} f)(\boldsymbol{R}(t_m)) (t_{m+1} - t_m)$$

$$+ \sqrt{2\kappa} \sum_{m=0}^{n} [\boldsymbol{\beta}(t_{m+1}) - \boldsymbol{\beta}(t_m)] \cdot (\boldsymbol{\nabla} f)(\boldsymbol{R}(t_m))$$

$$+ \kappa \sum_{m=0}^{n} [\beta^i(t_{m+1}) - \beta^i(t_m)][\beta^j(t_{m+1}) - \beta^j(t_m)] \nabla_i \nabla_j f(\boldsymbol{R}(t_m))$$

$$+ o(\Delta t). \tag{2.82}$$

In the limit when $n \to \infty$ and $\Delta t \to 0$, the first sum on the right hand side gives the well defined integral

$$\int_{t_0}^{t} \boldsymbol{v}(s, \boldsymbol{R}(s)) \cdot \boldsymbol{\nabla} f(\boldsymbol{R}(s)) \, \mathrm{d}s.$$

The second sum has also the form of a Riemann sum approximation to an integral. It converges when $n \to \infty$ and $\Delta t \to 0$. By definition, its limit gives the **Itô stochastic integral**

$$\int_{t_0}^{t} (\boldsymbol{\nabla} f)(\boldsymbol{R}(s)) \cdot \mathrm{d}\boldsymbol{\beta}(s).$$

Note that in the Riemann sum for it, the increments of the Brownian motion, that have zero mean, refer to later times than those involved in the prefactors. This is the essence of the Itô convention. As the result, the Itô stochastic integral has vanishing mean. Finally, the third sum on the right hand side of eq. (2.82) tends when $n \to \infty$ and $\Delta t \to 0$, essentially by the Law of Large Numbers, to the limit of its expectation value, i.e. to

$$\int_{t_0}^{t} (\boldsymbol{\nabla}^2 f)(\boldsymbol{R}(s)) \, \mathrm{d}s,$$

implementing the mnemotechnical rule of the **Itô calculus**

$$\mathrm{d}\beta^i(t) \, \mathrm{d}\beta^j(t) = \delta^{ij} |\mathrm{d}t|.$$

Altogether, we obtain the Itô stochastic integral equation

$$f(\boldsymbol{R}(t)) - f(\boldsymbol{R}(t_0)) = \int_{t_0}^{t} \boldsymbol{v}(s, \boldsymbol{R}(s)) \cdot \boldsymbol{\nabla} f(\boldsymbol{R}(s)) \, \mathrm{d}s$$

$$+ \sqrt{2\kappa} \int_{t_0}^{t} (\boldsymbol{\nabla} f)(\boldsymbol{R}(s)) \cdot \mathrm{d}\boldsymbol{\beta}(s) + \kappa \int_{t_0}^{t} (\boldsymbol{\nabla}^2 f)(\boldsymbol{R}(s)) \, \mathrm{d}s \tag{2.83}$$

which is equivalent to the Itô SDE (2.18). For the expectation values, it
gives the integral relation

$$\overline{f(\boldsymbol{R}(t)) - f(\boldsymbol{R}(t_0))} = \int_{t_0}^{t} \overline{[\boldsymbol{v}(s, \boldsymbol{R}(s)) \cdot \boldsymbol{\nabla} f(\boldsymbol{R}(s)) + \kappa (\boldsymbol{\nabla}^2 f)(\boldsymbol{R}(s))]} \, \mathrm{d}s$$

or the differential equation

$$\frac{\mathrm{d}}{\mathrm{d}t} \overline{f(\boldsymbol{R})} = \overline{\boldsymbol{v}(t, \boldsymbol{R}) \cdot \boldsymbol{\nabla} f(\boldsymbol{R})} + \overline{\kappa (\boldsymbol{\nabla}^2 f)(\boldsymbol{R})}.$$

In other words, the term proportional to $\mathrm{d}\boldsymbol{\beta}(t)$ in the Itô SDE does not
contribute to the equation for the mean values.

The **Stratonovich stochastic integral** is defined as the limit of an approximating Riemann sum where the prefactors of the Brownian increments
are taken at the middle point of the time interval. In our case

$$\int_{t_0}^{t} (\boldsymbol{\nabla} f)(\boldsymbol{R}(s)) \cdot \circ \, \mathrm{d}\boldsymbol{\beta}(s) = \lim_{n \to \infty} \sum_{m=0}^{n} (\boldsymbol{\nabla} f)(\boldsymbol{R}(\tfrac{1}{2}[t_m + t_{m+1}]))$$
$$\cdot [\boldsymbol{\beta}(t_{m+1}) - \boldsymbol{\beta}(t_m)], \quad (2.84)$$

where the circle in front of $d\boldsymbol{\beta}$ is the standard notation to distinguish the
Stratonovich convention from the Itô one. To compare to the Itô integral,
we expand

$$(\nabla_j f)(\boldsymbol{R}(\tfrac{1}{2}[t_m + t_{m+1}])) = (\nabla_j f)(\boldsymbol{R}(t_m))$$
$$+ \sqrt{2\kappa} (\nabla_i \nabla_j f)(\boldsymbol{R}(t_m)) \, [\beta^i(\tfrac{1}{2}[t_m + t_{m+1}]) - \beta^i(t_m)] + o((\Delta t)^{\frac{1}{2}})$$

and, upon substitution into the Riemann sum in (2.84), we obtain the relation

$$\int_{t_0}^{t} (\boldsymbol{\nabla} f)(\boldsymbol{R}(s)) \cdot \circ \, \mathrm{d}\boldsymbol{\beta}(s) = \int_{t_0}^{t} (\boldsymbol{\nabla} f)(\boldsymbol{R}(s)) \cdot \mathrm{d}\boldsymbol{\beta}(s)$$
$$+ \sqrt{2\kappa} \lim_{n \to \infty} \sum_{m=0}^{n} (\nabla_i \nabla_j f)(\boldsymbol{R}(t_m)) \, [\beta^j(t_{m+1}) - \beta^j(t_m)]$$
$$\cdot [\beta^i(\tfrac{1}{2}[t_m + t_{m+1}]) - \beta^i(t_m)]$$
$$= \int_{t_0}^{t} (\boldsymbol{\nabla} f)(\boldsymbol{R}(s)) \cdot \mathrm{d}\boldsymbol{\beta}(s) + \sqrt{\frac{\kappa}{2}} \int_{t_0}^{t} (\boldsymbol{\nabla}^2 f)(\boldsymbol{R}(s)) \, \mathrm{d}s,$$

where the last equality is again a consequence of the Law of Large Numbers. Note that, in general, the mean of the Stratonovich stochastic integral
does not vanish, which is a drawback of this convention. Comparison with

eq. (2.83) gives the Stratonovich integral equation

$$f(\boldsymbol{R}(t)) - f(\boldsymbol{R}(t_0)) = \int_{t_0}^t \boldsymbol{v}(s, \boldsymbol{R}(s)) \cdot \boldsymbol{\nabla} f(\boldsymbol{R}(s)) \, \mathrm{d}s$$
$$+ \sqrt{2\kappa} \int_{t_0}^t (\boldsymbol{\nabla} f)(\boldsymbol{R}(s)) \cdot \circ \, \mathrm{d}\boldsymbol{\beta}(s)$$

which is equivalent to the SDE (2.19). Observe, that the latter may be formally obtained from the SDE (2.17) by applying standard rules of differential calculus, unlike the Itô equation (2.18). This is the principal virtue of the Stratonovich convention.

Problem 1.2 As in the case without noise, we consider

$$\int f(\boldsymbol{r}) \, \partial_t n(t, \boldsymbol{r}) \, \mathrm{d}\boldsymbol{r} = \frac{\mathrm{d}}{\mathrm{d}t} \int f(\boldsymbol{r}) \, n(t, \boldsymbol{r}) \, \mathrm{d}\boldsymbol{r} .$$

Upon the substitution of eq. (2.20) for $n(t, \boldsymbol{r})$, the right hand side becomes

$$\frac{\mathrm{d}}{\mathrm{d}t} \int f(\boldsymbol{r}) \, \overline{\delta(\boldsymbol{r} - \boldsymbol{R}(t; t_0, \boldsymbol{r}_0))} \, n(t_0, \boldsymbol{r}_0) \, \mathrm{d}\boldsymbol{r}_0 \, \mathrm{d}\boldsymbol{r}$$
$$= \frac{\mathrm{d}}{\mathrm{d}t} \int \overline{f(\boldsymbol{R}(t; t_0, \boldsymbol{r}_0))} \, n(t_0, \boldsymbol{r}_0) \, \mathrm{d}\boldsymbol{r}_0 .$$

With the use of the Itô SDE (2.18), the last expression may be rewritten as

$$\int \overline{[(\boldsymbol{\nabla} f)(\boldsymbol{R}(t; t_0, \boldsymbol{r}_0)) \cdot \boldsymbol{v}(t, \boldsymbol{R}(t; t_0, \boldsymbol{r}_0)) + \kappa (\boldsymbol{\nabla}^2 f)(\boldsymbol{R}(t; t_0, \boldsymbol{r}_0))]} \, n(t_0, \boldsymbol{r}_0) \, \mathrm{d}\boldsymbol{r}_0$$

$$= \int [(\boldsymbol{\nabla} f)(\boldsymbol{r}) \cdot \boldsymbol{v}(t, \boldsymbol{r}) + \kappa (\boldsymbol{\nabla}^2 f)(\boldsymbol{r})] \, \overline{\delta(\boldsymbol{r} - \boldsymbol{R}(t; t_0, \boldsymbol{r}_0))} \, n(t_0, \boldsymbol{r}_0) \, \mathrm{d}\boldsymbol{r} \, \mathrm{d}\boldsymbol{r}_0 .$$

Upon using again eq. (2.20), this becomes

$$\int [(\boldsymbol{\nabla} f)(\boldsymbol{r}) \cdot \boldsymbol{v}(t, \boldsymbol{r}) + \kappa (\boldsymbol{\nabla}^2 f)(\boldsymbol{r})] \, n(t, \boldsymbol{r}) \, \mathrm{d}\boldsymbol{r}$$
$$= \int f(\boldsymbol{r}) [-\boldsymbol{\nabla} \cdot (n(t, \boldsymbol{r}) \boldsymbol{v}(t, \boldsymbol{r})) + \kappa \boldsymbol{\nabla}^2 n(t, \boldsymbol{r})] \, \mathrm{d}\boldsymbol{r} .$$

Stripping the result of the integral with an arbitrary function f, we obtain the desired equation for $n(t, \boldsymbol{r})$.

Problem 1.3 Suppose that $\theta(t, \boldsymbol{r})$ solves the evolution equation (2.6) with $g = 0$. To demonstrate that the relation (2.21) holds, it is enough to show that the derivative of the right hand side over t_0 vanishes, because the relation holds trivially at $t = t_0$. By the Itô calculus,

$$\frac{\mathrm{d}}{\mathrm{d}t_0} \overline{\theta(t_0, \boldsymbol{R}(t_0; t, \boldsymbol{r}))} = \overline{\partial_{t_0} \theta(t_0, \boldsymbol{R}(t_0; t, \boldsymbol{r}))}$$

$$+ \overline{v(t_0, R(t_0; t, r)) \cdot (\nabla \theta)(t_0, R(t_0; t, r))} - \kappa \overline{(\nabla^2 \theta)(t_0, R(t_0; t, r))} = 0.$$

Note the minus sign in front of the $\kappa \nabla^2$ term. Where does it come from? The last equality follows from the assumed equation for $\theta(t, r)$.

Problem 1.4 For a vector-valued function f,

$$\int f(r) \cdot \partial_t B(t, r) \, dr = \frac{d}{dt} \int f(r) \cdot B(t, r) \, dr$$

$$= \frac{d}{dt} \int \overline{f_i(R(t; t_0, r_0)) \, W^i{}_j(t; t_0, r_0)} \, B^j(t_0, r_0) \, dr_0,$$

where we have substituted the assumed relation (2.22) for $B(t, r)$ and performed the integral over r. With the use of the Itô SDE (2.18) and of eq. (2.14), the right hand side may be rewritten as

$$= \int \overline{v(t, R(t; t_0, r_0)) \cdot (\nabla f_i)(R(t; t_0, r_0)) \, W^i{}_j(t; t_0, r_0)} \, B^j(t_0, r_0) \, dr_0$$

$$+ \kappa \int \overline{(\nabla^2 f_i)(R(t; t_0, r_0)) \, W^i{}_j(t; t_0, r_0)} \, B^j(t_0, r_0) \, dr_0$$

$$+ \int \overline{f_i(R(t; t_0, r_0)) \, (\nabla_k v^i)(t, R(t; t_0, r_0)) \, W^k{}_j(t; t_0, r_0)} \, B^j(t_0, r_0) \, dr_0$$

$$= \int [v(t, r) \cdot (\nabla f_i)(r) + \kappa(\nabla^2 f_i)(r) + f(r) \cdot (\nabla_i v)(t, r)]$$

$$\cdot \overline{\delta(r - R(t; t_0, r_0)) \, W^i{}_j(t; t_0, r_0)} \, B^j(t_0, r_0) \, dr \, dr_0.$$

Because of eq. (2.14), the last expression is equal to

$$\int [v(t, r) \cdot (\nabla f_i)(r) + \kappa(\nabla^2 f_i)(r) + f(r) \cdot (\nabla_i v)(t, r)] \, B^i(t, r) \, dr$$

$$= \int f_i(r) [-v(t, r) \cdot \nabla B^i(t, r) + B(t, r) \cdot \nabla v^i(t, r)$$

$$+ (\nabla \cdot v(t, r)) \, B^i(t, r) + \kappa \nabla^2 B^i(t, r)] \, dr.$$

Stripping the obtained identity from the integral against an arbitrary function f leads to the desired equation for B.

2.9.2 Problems to Lecture 2

Problem 2.1 By definition (2.26) of Φ_t, we have

$$\Phi_t \circ \Phi_s(r, v) = \Phi_t(R(s; r|v), v_s) = (R(t; R(s; r|v)|v_s), v_{t+s}). \quad (2.85)$$

Since

$$R(t; R(s; r|v)|v_s) = R(t + s; r|v),$$

as both sides satisfy the same differential equation in t and take the same value at $t = s$, we infer that the right hand side of eq. (2.85) is equal to $\Phi_{t+s}(\boldsymbol{r}, \boldsymbol{v})$.

Problem 2.2 Using the invariance of the measure N, see eq. (2.28), we infer that

$$P_{-t}(\vec{\rho}) = \int \delta(\vec{\rho} - \vec{\rho}(-t; \boldsymbol{r}|\boldsymbol{v}))\, N(\mathrm{d}\boldsymbol{r}, \mathrm{d}\boldsymbol{v})$$

$$= \int \delta(\vec{\rho} - \vec{\rho}(-t; \boldsymbol{R}(t; \boldsymbol{r}|\boldsymbol{v})|\boldsymbol{v}_t))\, N(\mathrm{d}\boldsymbol{r}, \mathrm{d}\boldsymbol{v})\,.$$

By virtue of eq. (2.31), the last integral is equal to

$$\int \delta(\vec{\rho} + \bar{\rho}(t, \boldsymbol{r}|\boldsymbol{v}))\, N(\mathrm{d}\boldsymbol{r}, \mathrm{d}\boldsymbol{v})$$

$$= \int \delta(-\vec{\rho} - \bar{\rho}(t, \boldsymbol{r}|\boldsymbol{v}))\, N(\mathrm{d}\boldsymbol{r}, \mathrm{d}\boldsymbol{v}) = P_t(-\vec{\rho})\,.$$

This establishes the identity (2.36).

Problem 2.3 Although one expects that for large positive times, the difference between $P_t(\vec{\rho})$ and $\tilde{P}_t(\vec{\rho})$ disappears since the forward stretching exponents will forget whether their initial points were sampled with the natural measures determined by past velocities or with the uniform measure, there is no reason to expect such asymptotic identity for large negative times since, in general compressible flows, the backward stretching exponents are strongly correlated with the natural time zero measures. Indeed, backward trajectories with initial points sampled with these measures are very atypical in compressible flows. They stay on the attractors for the forward evolution whereas typical backward trajectories approach attractors for the backward evolution. In particular, in time reversible velocities, there is a (phase-)space expansion around backward trajectories with average rate equal to $-\sum \lambda_i$ when the initial points are sampled with natural measures, whereas there is a (phase-)space contraction around backward trajectories with the same average rate equal to $-\sum \lambda_i$ if the initial points are sampled uniformly.

2.9.3 Problems to Lecture 3

Problem 3.1 The integral equation corresponding to the SDE (2.37) is

$$\boldsymbol{R}(t) - \boldsymbol{R}(t_0) = \int_{t_0}^{t} \mathrm{d}\boldsymbol{w}(s, \boldsymbol{R}(s)) = \int_{t_0}^{t} \boldsymbol{v}(s, \boldsymbol{R}(s))\, \mathrm{d}s\,.$$

With the Itô convention,

$$\int_{t_0}^t v(s, R(s))\, ds = \lim_{n\to\infty} \sum_{m=0}^n \int_{t_m}^{t_{m+1}} v(s, R(t_m))\, ds \equiv \mathcal{I}.$$

With the Stratonovich convention,

$$\int_{t_0}^t v(s, R(s))\, ds = \lim_{n\to\infty} \sum_{m=0}^n \int_{t_m}^{t_{m+1}} v(s, R(\tfrac{1}{2}[t_m + t_{m+1}]))\, ds \equiv \mathcal{S}.$$

For the difference, we get

$$
\begin{aligned}
\mathcal{S} - \mathcal{I} &= \lim_{n\to\infty} \sum_{m=0}^n \int_{t_m}^{t_{m+1}} \left[v(s, R(\tfrac{1}{2}[t_m + t_{m+1}])) - v(s, R(t_m)) \right]\, ds \\
&= \lim_{n\to\infty} \sum_{m=0}^n \int_{t_m}^{t_{m+1}} \left[R^i(\tfrac{1}{2}[t_m + t_{m+1}]) - R^i(t_m) \right] \nabla_i v(s, R(t_m))\, ds \\
&= \lim_{n\to\infty} \sum_{m=0}^n \int_{t_m}^{\frac{1}{2}[t_m + t_{m+1}]} v^i(s', R(t_m))\, ds' \int_{t_m}^{t_{m+1}} \nabla_i v(s, R(t_m))\, ds \\
&= -\tfrac{1}{2}(t - t_0) \nabla_i D^{i\cdot}(\mathbf{0}).
\end{aligned}
$$

because in the limit the sum tends to its average by the Law of Large Numbers. Due to isotropy, $\nabla_i D^{ij}(\mathbf{0}) = 0$ so that in homogeneous isotropic Kraichnan velocities both conventions for the SDE (2.37) coincide.

Problem 3.2 The integral equation equivalent to the SDE (2.40) is

$$\delta R(t) - \delta R(t_0) = \int_{t_0}^t (\delta R(s) \cdot \nabla) v(s, R(s))\, ds.$$

With the Itô convention,

$$
\begin{aligned}
&\int_{t_0}^t (\delta R(s) \cdot \nabla) v(s, R(s))\, ds \\
&\quad = \lim_{n\to\infty} \sum_{m=0}^n \int_{t_m}^{t_{m+1}} (\delta R(t_m) \cdot \nabla)\, v(s, R(t_m))\, ds \equiv \mathcal{I}
\end{aligned}
$$

and with the Stratonovich convention,

$$
\begin{aligned}
&\int_{t_0}^t (\delta R(s) \cdot \nabla) \circ v(s, R(s))\, ds \\
&\quad = \lim_{n\to\infty} \sum_{m=0}^n \int_{t_m}^{t_{m+1}} (\delta R(\tfrac{1}{2}[t_m + t_{m+1}]) \cdot \nabla)\, v(s, R(\tfrac{1}{2}[t_m + t_{m+1}]))\, ds \equiv \mathcal{S}
\end{aligned}
$$

The difference is

$$\mathcal{S} - \mathcal{I}$$

$$= \lim_{n\to\infty} \sum_{m=0}^{n} \int_{t_m}^{t_{m+1}} [\delta \boldsymbol{R}(\tfrac{1}{2}[t_m + t_{m+1}]) - \delta \boldsymbol{R}(t_m)] \cdot \boldsymbol{\nabla}) \, \boldsymbol{v}(s, \boldsymbol{R}(\tfrac{1}{2}[t_m + t_{m+1}])) \, \mathrm{d}s$$

$$+ \lim_{n\to\infty} \sum_{m=0}^{n} \int_{t_m}^{t_{m+1}} (\delta \boldsymbol{R}(t_m) \cdot \boldsymbol{\nabla})[\boldsymbol{v}(s, \boldsymbol{R}(\tfrac{1}{2}[t_m + t_{m+1}])) - \boldsymbol{v}(s, \boldsymbol{R}(t_m))] \, \mathrm{d}s \,.$$

In the first integral, $\boldsymbol{v}(s, \boldsymbol{R}(\tfrac{1}{2}[t_m + t_{m+1}]))$ may be replaced by $\boldsymbol{v}(s, \boldsymbol{R}(t_m))$ because the error disappears in the limit. Similarly, in the second integral, it is enough to expand the difference $\boldsymbol{v}(s, \boldsymbol{R}(\tfrac{1}{2}[t_m + t_{m+1}])) - \boldsymbol{v}(s, \boldsymbol{R}(t_m))$ to the first order in $\boldsymbol{R}(\tfrac{1}{2}[t_m + t_{m+1}]) - \boldsymbol{R}(t_m)$. This gives

$$\mathcal{S} - \mathcal{I} = \lim_{n\to\infty} \sum_{m=0}^{n} \int_{t_m}^{t_{m+1}} [\delta \boldsymbol{R}(\tfrac{1}{2}[t_m + t_{m+1}]) - \delta \boldsymbol{R}(t_m)] \cdot \boldsymbol{\nabla}) \, \boldsymbol{v}(s, \boldsymbol{R}(t_m)) \, \mathrm{d}s$$

$$+ \lim_{n\to\infty} \sum_{m=0}^{n} \int_{t_m}^{t_{m+1}} [R^i(\tfrac{1}{2}[t_m + t_{m+1}]) - R^i(t_m)] \, (\delta \boldsymbol{R}(t_m) \cdot \nabla_i \boldsymbol{\nabla}) \, \boldsymbol{v}(s, \boldsymbol{R}(t_m)) \, \mathrm{d}s$$

$$= \lim_{n\to\infty} \sum_{m=0}^{n} \int_{t_m}^{\frac{1}{2}[t_m + t_{m+1}]} [\delta \boldsymbol{R}(t_m) \cdot \boldsymbol{\nabla} v^i](s', \boldsymbol{R}(t_m)) \, \mathrm{d}s' \int_{t_m}^{t_{m+1}} \nabla_i \, \boldsymbol{v}(s, \boldsymbol{R}(t_m)) \, \mathrm{d}s$$

$$+ \lim_{n\to\infty} \sum_{m=0}^{n} \int_{t_m}^{\frac{1}{2}[t_m + t_{m+1}]} v^i(s', \boldsymbol{R}(t_m)) \, \mathrm{d}s' \int_{t_m}^{t_{m+1}} (\delta \boldsymbol{R}(t_m) \cdot \nabla_i \boldsymbol{\nabla}) \, \boldsymbol{v}(s, \boldsymbol{R}(t_m)) \, \mathrm{d}s$$

$$= \tfrac{1}{2} \int_{t_0}^{t} \left[-(\delta \boldsymbol{R}(s) \cdot \boldsymbol{\nabla}) \nabla_i D^{i\,\cdot}(\boldsymbol{0}) + (\delta \boldsymbol{R}(s) \cdot \nabla_i \boldsymbol{\nabla}) D^{i\,\cdot}(\boldsymbol{0}) \right] \; = \; 0 \,.$$

Problem 3.3 The integral equation corresponding to the SDE (2.42) is

$$\delta \boldsymbol{R}(t) - \delta \boldsymbol{R}(t_0) \; = \; \int_{t_0}^{t} S(t) \mathrm{d}s \, \delta \boldsymbol{R}(s) \,.$$

With the Itô convention,

$$\int_{t_0}^{t} S(t) \mathrm{d}s \, \delta \boldsymbol{R}(s) \; = \; \lim_{n\to\infty} \sum_{m=0}^{n} \int_{t_m}^{t_{m+1}} S(s) \mathrm{d}s \, \delta \boldsymbol{R}(t_m) \; \equiv \mathcal{I} \,,$$

whereas with the Stratonovich one,

$$\int_{t_0}^{t} S(t) \mathrm{d}s \circ \delta \boldsymbol{R}(s) \; = \; \lim_{n\to\infty} \sum_{m=0}^{n} \int_{t_m}^{t_{m+1}} S(s) \mathrm{d}s \, \delta \boldsymbol{R}(\tfrac{1}{2}[t_m + t_{m+1}]) \; \equiv \mathcal{S}$$

so that

$$\boldsymbol{\mathcal{S}} - \boldsymbol{\mathcal{I}} = \lim_{n\to\infty} \sum_{m=0}^{n} \int_{t_m}^{t_{m+1}} S(s)\mathrm{d}s\, [\delta\boldsymbol{R}(\tfrac{1}{2}[t_m + t_{m+1}]) - \delta\boldsymbol{R}(t_m)]$$

$$= \lim_{n\to\infty} \sum_{m=0}^{n} \int_{t_m}^{t_{m+1}} S(s)\mathrm{d}s \int_{t_m}^{\frac{1}{2}[t_m + t_{m+1}]} S(s')\,\delta\boldsymbol{R}(t_m)]\,\mathrm{d}s'$$

$$= \tfrac{1}{2} \int_{t_0}^{t} C_{jl}^{\cdot j}\, \delta R^l(s)\,\mathrm{d}s = -\tfrac{1}{2} \int_{t_0}^{t} (\delta\boldsymbol{R}(s) \cdot \boldsymbol{\nabla})\nabla_i D^{i\cdot}(\boldsymbol{0})\,\mathrm{d}s\,.$$

According to eqs. (2.45) and (2.46),

$$C_{jl}^{ij} = ((d+1)\beta + \gamma)\,\delta_l^i = (2\beta + d\gamma)\wp\,\delta_l^i$$

so that the choice of the Itô or Stratonovich convention does matter here in the presence of compressibility. Note that the difference between the result when $S^i{}_j(t)$ is the white noise and when it is replaced by $\Sigma^i{}_j(t) = \nabla_j v^i(t, \boldsymbol{R}(t))$ came from the t-dependence of $\boldsymbol{R}(t)$ in the last formula.

Problem 3.4 In virtue of eqs. (2.47) and (2.45), we have

$$\mathscr{L} = \tfrac{1}{2} C_{kl}^{ij}\, W^k{}_m W^l{}_n \frac{\partial}{\partial W^i{}_m} \frac{\partial}{\partial W^j{}_n}$$

$$= \tfrac{1}{2}\left(\beta\, W^i{}_m W^j{}_n + \beta\, W^j{}_m W^i{}_n + \gamma\,\delta^{ij} W^k{}_m W^k{}_n\right) \frac{\partial}{\partial W^i{}_m} \frac{\partial}{\partial W^j{}_n}$$

$$= \tfrac{1}{2}\left(\beta\, W^i{}_m \frac{\partial}{\partial W^i{}_m} W^j{}_n \frac{\partial}{\partial W^j{}_n} - \beta\, W^i{}_m \frac{\partial}{\partial W^i{}_m} + \beta\, W^j{}_m \frac{\partial}{\partial W^i{}_m} W^i{}_n \frac{\partial}{\partial W^j{}_n}\right.$$

$$\left. -d\,\beta\, W^j{}_m \frac{\partial}{\partial W^j{}_m} + \gamma\, W^k{}_m \frac{\partial}{\partial W^i{}_m} W^k{}_n \frac{\partial}{\partial W^i{}_n} - \gamma\, W^i{}_m \frac{\partial}{\partial W^i{}_m}\right)$$

Since

$$\mathscr{E}_i{}^j = -(E_i{}^j W)^n{}_m \frac{\partial}{\partial W^n{}_m} = -\delta_i^n \delta_r^j\, W^r{}_m \frac{\partial}{\partial W^n{}_m} = -W^j{}_m \frac{\partial}{\partial W^i{}_m}$$

and

$$\mathscr{D} = W^i{}_m \frac{\partial}{\partial W^i{}_m}\,,$$

the desired identity (2.54) follows.

Problem 3.5 First, let us observe that for $\mathscr{J}_{ij} = \mathscr{E}_i{}^j - \mathscr{E}_j{}^i$,

$$(\mathscr{J}_{ij}f)(W) = \frac{\mathrm{d}}{\mathrm{d}\epsilon}\Big|_0 f(e^{-\epsilon(E_i{}^j - E_j{}^i)}W) \tag{2.86}$$

vanishes when acting on left $O(d)$-invariant functions, so that on such functions,

$$\mathscr{L} = \frac{\beta+\gamma}{2} \sum_{i,j} \mathscr{E}_i{}^j \mathscr{E}_j{}^i + \frac{\beta}{2} \mathscr{D}^2 - \frac{(d+1)\beta+\gamma}{2} \mathscr{D}.$$

In order to compute $(\mathcal{L}f)(\vec{\rho})$, it will be enough to compute $(\mathscr{L}f)(W_\rho)$ for left-right $O(d)$-invariant functions f and W_ρ a diagonal matrix with the entries $e^{\rho_1}, \ldots, e^{\rho_d}$. Clearly,

$$(\mathscr{E}_i{}^i f)(W_\rho) = -\frac{\partial}{\partial \rho_i} f(\vec{\rho}), \qquad ((\mathscr{E}_i{}^i)^2 f)(W_\rho) = \frac{\partial^2}{\partial \rho_i^2} f(\vec{\rho}),$$

$$(\mathscr{D}f)(W_\rho) = \sum_i \frac{\partial}{\partial \rho_i} f(\vec{\rho}), \qquad (\mathscr{D}f)(W_\rho) = \Big(\sum_i \frac{\partial}{\partial \rho_i}\Big)^2 f(\vec{\rho})$$

so that

$$\left[\frac{\beta+\gamma}{2} \sum_i \mathscr{E}_i{}^i \mathscr{E}_i{}^i + \frac{\beta}{2} \mathscr{D}^2 - \frac{(d+1)\beta+\gamma}{2} \mathscr{D}\right] f(W_\rho)$$

$$= \left[\frac{\beta+\gamma}{2} \sum_i \frac{\partial^2}{\partial \rho_i^2} + \frac{\beta}{2} \Big(\sum_i \frac{\partial}{\partial \rho_i}\Big)^2 - \frac{(d+1)\beta+\gamma}{2} \Big(\sum_i \frac{\partial}{\partial \rho_i}\Big)\right] f(\vec{\rho}) . \quad (2.87)$$

We are left with calculating

$$\left[\frac{\beta+\gamma}{2} \sum_{i\neq j} \mathscr{E}_i{}^j \mathscr{E}_j{}^i\right] f(W_\rho).$$

Since, for $i \neq j$, $\mathscr{E}_i{}^j \mathscr{E}_j{}^i$ acts only on the 2×2 sub-matrix corresponding to the indices i and j, it is enough to do the calculations of $\mathscr{E}_1{}^2 \mathscr{E}_2{}^1 + \mathscr{E}_2{}^1 \mathscr{E}_1{}^2$ in two dimensions. We have,

$$(\mathscr{E}_1{}^2 \mathscr{E}_2{}^1 f)\begin{pmatrix} e^{\rho_1} & 0 \\ 0 & e^{\rho_2} \end{pmatrix} = \frac{\mathrm{d}}{\mathrm{d}\epsilon_1} \frac{\mathrm{d}}{\mathrm{d}\epsilon_2}\Big|_0 f\Big(e^{-\epsilon_2 E_2{}^1} e^{-\epsilon_1 E_1{}^2} \begin{pmatrix} e^{\rho_1} & 0 \\ 0 & e^{\rho_2} \end{pmatrix}\Big)$$

$$= \frac{\mathrm{d}}{\mathrm{d}\epsilon_1} \frac{\mathrm{d}}{\mathrm{d}\epsilon_2}\Big|_0 f(W_\epsilon),$$

where

$$W_\epsilon = \begin{pmatrix} e_1^\rho & -\epsilon_1 e^{\rho_2} \\ -\epsilon_2 e^{\rho_1} & (1+\epsilon_1\epsilon_2) e^{\rho_2} \end{pmatrix}$$

Since $f(W_\epsilon)$ depends only on the stretching exponents of W_ϵ, we have to find them, keeping at the end only terms proportional to $\epsilon_1\epsilon_2$. The eigenvalues of the matrix

$$W_\epsilon W_\epsilon^T = \begin{pmatrix} e^{2\rho_1} & -\epsilon_2 e^{2\rho_1} - \epsilon_1 e^{2\rho_2} \\ -\epsilon_2 e^{2\rho_1} - \epsilon_1 e^{2\rho_2} & (1+2\epsilon_1\epsilon_2) e^{2\rho_2} \end{pmatrix} + \cdots,$$

satisfy the equation

$$\lambda^2 - \lambda \left(e^{2\rho_1} + e^{2\rho_2} + 2\epsilon_1\epsilon_2 e^{2(\rho_1+\rho_2)}\right) + e^{2(\rho_1+\rho_2)} + \cdots = 0$$

from which we obtain

$$\lambda = \begin{cases} e^{2\rho_1}\left(1 + 2\epsilon_1\epsilon_2 \dfrac{e^{2\rho_2}}{e^{2\rho_1}-e^{2\rho_2}}\right) + \cdots , \\ e^{2\rho_2}\left(1 - 2\epsilon_1\epsilon_2 \dfrac{e^{2\rho_2}}{e^{2\rho_1}-e^{2\rho_2}}\right) + \cdots , \end{cases}$$

so that the modified stretching exponents are

$$\rho_1 + 2\epsilon_1\epsilon_2 \frac{e^{2\rho_2}}{e^{2\rho_1} - e^{2\rho_2}} + \cdots$$

$$\rho_2 - \epsilon_1\epsilon_2 \frac{e^{2\rho_2}}{e^{2\rho_1} - e^{2\rho_2}} + \cdots$$

We infer that

$$\left(\mathscr{E}_1{}^2\mathscr{E}_2{}^1 f\right)\begin{pmatrix} e^{\rho_1} & 0 \\ 0 & e^{\rho_2} \end{pmatrix} = \frac{e^{2\rho_2}}{e^{2\rho_1} - e^{2\rho_2}}\left(\frac{\partial}{\partial\rho_1} - \frac{\partial}{\partial\rho_2}\right) f(\rho_1, \rho_2) .$$

Similar calculation shows that

$$\left(\mathscr{E}_2{}^1\mathscr{E}_1{}^2 f\right)\begin{pmatrix} e^{\rho_1} & 0 \\ 0 & e^{\rho_2} \end{pmatrix} = \frac{e^{2\rho_1}}{e^{2\rho_1} - e^{2\rho_2}}\left(\frac{\partial}{\partial\rho_1} - \frac{\partial}{\partial\rho_2}\right) f(\rho_1, \rho_2)$$

so that

$$\left[\frac{\beta+\gamma}{2}\sum_{i\neq j}\mathscr{E}_i{}^j\mathscr{E}_j{}^i f\right](W_\rho) = \frac{\beta+\gamma}{2}\sum_{i\neq j}\coth(\rho_i - \rho_j)\frac{\partial}{\partial\rho_i} f(\vec{\rho}) \qquad (2.88)$$

Adding eqs. (2.87) and (2.88) establishes formula (2.50).

Problem 3.6   First, setting $\rho_i - \rho_j \equiv \rho_{ij}$, note that

$$\mathcal{F}\frac{\partial}{\partial\rho_i}\mathcal{F}^{-1} = \frac{\partial}{\partial\rho_i} + \frac{1}{2} - \frac{1}{2}\sum_{j\neq i}\coth(\rho_{ij}) ,$$

$$\mathcal{F}\sum_i\frac{\partial}{\partial\rho_i}\mathcal{F}^{-1} = \sum_i\frac{\partial}{\partial\rho_i} + \frac{d}{2} .$$

Hence

$$\mathcal{F}\mathcal{L}\mathcal{F}^{-1} = \frac{\beta+\gamma}{2}\sum_i\left(\frac{\partial}{\partial\rho_i} + \frac{1}{2} - \frac{1}{2}\sum_{j\neq i}\coth(\rho_{ij})\right)^2$$

$$+ \frac{\beta+\gamma}{2}\sum_{i\neq j}\coth(\rho_{ij})\left(\frac{\partial}{\partial\rho_i} + \frac{1}{2} - \frac{1}{2}\sum_{k\neq i}\coth(\rho_{ik})\right)$$

$$+ \frac{\beta}{2}\left(\sum_i\frac{\partial}{\partial\rho_i} + \frac{d}{2}\right)^2 - \frac{(d+1)\beta+\gamma}{2}\left(\sum_i\frac{\partial}{\partial\rho_i} + \frac{d}{2}\right)$$

$$= \frac{\beta + \gamma}{2} \left(\sum_i \frac{\partial^2}{\partial \rho_i^2} + \frac{1}{2} \sum_{i \neq j} \frac{1}{\sinh^2(\rho_{ij})} - \frac{1}{4} \sum_i \sum_{\substack{j \neq i \\ k \neq i}} \coth(\rho_{ij}) \coth(\rho_{ik}) \right)$$

$$+ \frac{\beta}{2} \left(\sum_i \frac{\partial}{\partial \rho_i} \right)^2 - \frac{[(d+1)\beta + \gamma]d}{8}. \tag{2.89}$$

To simplify further the obtained expression, we note that

$$\sum_i \sum_{j \neq i} \sum_{k \neq i} \coth(\rho_{ij}) \coth(\rho_{ik}) = \sum_{i \neq j} \coth^2(\rho_{ij})$$

$$+ 2 \sum_{i<j<k} \left[\coth(\rho_{ij}) \coth(\rho_{ik}) + \coth(\rho_{ki}) \coth(\rho_{kj}) + \coth(\rho_{jk}) \coth(\rho_{ji}) \right]$$

$$= \sum_{i \neq j} \frac{1}{\sinh^2(\rho_{ij})} + \frac{(d^2-1)d}{3}$$

since the expression in the brackets is equal to 1 (check it!). Upon the substitution of the last relation to eq. (2.89), the identity (2.51) follows with

$$\text{const.} = \frac{[3\beta d + (\beta + \gamma)(d^2 + 2)]d}{24}.$$

This is the bottom of the spectrum of $\mathcal{H}_{CSM}$. Indeed, the first entry in this operator,

$$-\frac{\beta + \gamma}{2} \left(\sum_i \frac{\partial^2}{\partial \rho_i^2} + \frac{1}{2} \sum_{i<j} \frac{1}{\sinh^2(\rho_i - \rho_j)} \right)$$

has a continuous spectrum that extends down to zero and no bound states, in spite of the attractive potential [Ols83]. The second entry, the center-of-mass Laplacian

$$-\frac{\beta}{2} \left(\sum_i \frac{\partial}{\partial \rho_i} \right)^2,$$

commutes with the first one and has the same spectrum.

Problem 3.7 For quadratic $H(\vec{\sigma}) = \frac{1}{2} A^{ij} \sigma_i \sigma_j + B^i \sigma_i + C$, the relation MFR requires that

$$A^{(d+1-i)(d+1-j)} = A^{ij}, \qquad -B^{d+1-i} = B^i - 1. \tag{2.90}$$

For $H(\vec{\sigma})$ given by (2.52),

$$A^{ij} = \frac{1}{\beta + \gamma} \delta^{ij} - \frac{\beta}{(\beta + \gamma)((d+1)\beta + \gamma)},$$

$$B^i = -\frac{1}{\beta + \gamma} \left(\lambda_i - \frac{\beta}{(d+1)\beta + \gamma} \sum_j \lambda_j \right)$$

so that the first of conditions (2.90) for A^{ij} is satisfied and the second one for B^i becomes

$$B^i + B^{d+1-i} = -\frac{1}{\beta + \gamma}\Big(\lambda_{d+1-i} + \lambda_i - \frac{2\beta}{(d+1)\beta + \gamma}\sum_j \lambda_j\Big) = 1. \quad (2.91)$$

Since, in virtue of eq. (2.53)

$$\lambda_{d+1-i} + \lambda_i = -((d+1)\beta + \gamma) = \frac{2}{d}\sum_j \lambda_j,$$

the relation (2.91) readily follows. Note that the average space contraction rate $-\sum_j \lambda_j = \frac{1}{2}d(2\beta + d\gamma)\wp$ is non-negative and it vanishes if and only if the compressibility degree $\wp$ vanishes.

2.9.4 Problems to Lecture 4

Problem 4.1 For $\Delta(t) \equiv |\Delta\mathbf{R}(t)|$, we obtain from eq. (2.57), using the short-distance form of the difference $D^{ij}(\mathbf{0}) - D^{ij}(\Delta\mathbf{R})$,

$$\frac{d}{dt}\langle f(\Delta)\rangle = \frac{1}{2}\Big\langle \Delta R^k \Delta R^l C^{ij}_{kl} \nabla_i \nabla_j f(\Delta)\Big\rangle$$
$$= \frac{1}{2}\Big\langle \Delta R^k \Delta R^l C^{ij}_{kl} \frac{1}{\Delta^2}\Big(\delta^{ij}\frac{d}{d\ln\Delta}f(\Delta) - 2\Delta R^i \Delta R^j \frac{1}{\Delta^2}\frac{d}{d\ln\Delta}f(\Delta)$$
$$+ \Delta R^i \Delta R^j \frac{1}{\Delta^2}\Big(\frac{d}{d\ln\Delta}\Big)^2 f(\Delta)\Big)\Big\rangle.$$

With $C^{ij}_{kl} = -\nabla_k\nabla_l D^{ij}(\mathbf{0})$ given by (2.45), this becomes eq. (2.58). In order to prove eq. (2.59), note that

$$e^{\frac{\lambda_1}{2\beta+\gamma}\ln\Delta}\Big[\frac{2\beta+\gamma}{2}\Big(\frac{d}{d\ln\Delta}\Big)^2 + \lambda_1\frac{d}{d\ln\Delta}\Big] e^{-\frac{\lambda_1}{2\beta+\gamma}\ln\Delta}$$
$$= \frac{2\beta+\gamma}{2}\Big(\frac{d}{d\ln\Delta}\Big)^2 - \frac{\lambda_1^2}{2(2\beta+\gamma)}.$$

Assuming the initial condition $\Delta(0) = \Delta_0$, it follows that

$$P_t(\Delta; \Delta_0) = \langle\delta(\Delta - \Delta(t))\rangle = \langle\delta(\ln\Delta - \ln\Delta(t))\rangle\frac{1}{\Delta}$$
$$= e^{\frac{\lambda_1}{2\beta+\gamma}(\ln\Delta - \ln\Delta_0) - \frac{\lambda_1^2 t}{2(2\beta+\gamma)}} e^{t\frac{2\beta+\gamma}{2}\big(\frac{d}{d\ln\Delta}\big)^2}(\Delta_0; \Delta)\frac{1}{\Delta}$$
$$= \frac{1}{\sqrt{2\pi(2\beta+\gamma)t}} e^{\frac{\lambda_1}{2\beta+\gamma}(\ln\Delta - \ln\Delta_0) - \frac{\lambda_1^2 t}{2(2\beta+\gamma)} - \frac{(\ln\Delta - \ln\Delta_0)^2}{2(2\beta+\gamma)t}}\frac{1}{\Delta}$$

which gives the log-normal PDF of eq. (2.59).

Problem 4.2 To prove the convergence (2.60), note that for a test function $f(\Delta)$,

$$\int f(\Delta)\, P_t(\Delta; \Delta_0)\, d\Delta = \frac{1}{\sqrt{2\pi(2\beta+\gamma)t}} \int f(\Delta)\, e^{-\frac{(\ln\frac{\Delta}{\Delta_0}-\lambda_1 t)^2}{2(2\beta+\gamma)t}}\, \frac{d\Delta}{\Delta}$$

$$= \frac{1}{\sqrt{2\pi(2\beta+\gamma)t}} \int f(\Delta_0\, e^s)\, e^{-\frac{(s-\lambda_1 t)^2}{2(2\beta+\gamma)t}}\, ds \xrightarrow[\Delta_0 \to 0]{} f(0)$$

by the Dominated Convergence Theorem.

Problem 4.3 The stationary scalar 2-point function $F^{(2)}(\Delta)$ satisfies the relation

$$-(MF^{(2)})(\Delta) = \chi(\Delta),$$

see eq. (2.71). In the weakly compressible phase, $D_{\text{eff}} > 2$ or $a > 1$ and the general solution to this equation has the form

$$F^{(2)}(\Delta) = \frac{2}{2\beta+\gamma} \int_\Delta^\infty (\Delta')^{-a}\, d\Delta' \left(\int_0^{\Delta'} (\Delta'')^{a-\xi} \chi(\Delta'')\, d\Delta'' + C_1 \right) + C_2.$$

The correct solution is chosen by imposing the boundary conditions

$$\Delta^a \frac{d}{d\Delta} F^{(2)}(0) = 0, \qquad F^{(2)}(\infty) = 0,$$

which yield $C_1 = C_2 = 0$. Note that this implies that the mean scalar energy density $F^{(2)}(0)$ is finite and positive in this phase. Changing the order of integration, we may rewrite the solution with $C_1 = C_2 = 0$ as

$$F^{(2)}(\Delta) = \int_0^\infty (-M)^{-1}(\Delta; \Delta')\, \chi(\Delta')\, d\Delta',$$

where

$$(-M)^{-1}(\Delta; \Delta') = \frac{2}{(2\beta+\gamma)(a-1)} \begin{cases} (\Delta')^{1-\xi} & \text{for } \Delta \le \Delta', \\ \Delta^{1-a}(\Delta')^{a-\xi} & \text{for } \Delta \ge \Delta' \end{cases}$$

is the Green function of the operator $-M$ with the reflecting boundary condition (2.64) at $\Delta = 0$. It is built by gluing two zero modes of the operator M, a constant and Δ^{1-a}, the first one more regular around $\Delta = 0$, the second one more regular around $\Delta = \infty$.

For the 2-point structure function $S^{(2)}(\Delta) = 2(F^{(2)}(0) - F^{(2)}(\Delta))$, we obtain

$$S^{(2)}(\Delta) = \frac{4}{2\beta+\gamma} \int_0^\Delta (\Delta')^{-a}\, d\Delta' \int_0^{\Delta'} (\Delta'')^{a-\xi} \chi(\Delta'')\, d\Delta''.$$

For small Δ, this scales as $\Delta^{2-\xi}$. The last formula holds also in the intermediate compressibility phase with the choice of the reflecting boundary condition (physically, for very small diffusivity and vanishing viscosity) although the 2-point function $F_t^{(2)}$ does not reach here the stationary form [Gaw00].

In the strongly compressible phase with $D_{\text{eff}} < 0$ or $a < \xi - 1$, the stationary 2-point structure function $S^{(2)}(\Delta)$ satisfies the relation

$$-MS^{(2)}(\Delta) = 2(\chi(0) - \chi(\Delta)),$$

see eq. (2.73). The correct solution has the form

$$
\begin{aligned}
S^{(2)}(\Delta) &= \frac{4}{2\beta + \gamma} \int_0^\Delta (\Delta')^{-a}\, d\Delta' \int_{\Delta'}^\infty (\Delta'')^{a-\xi}(\chi(0) - \chi(\Delta))\, d\Delta'' \\
&= \int_0^\infty (-M)^{-1}\, 2(\chi(0) - \chi(\Delta))\, d\Delta,
\end{aligned}
$$

where now

$$
(-M)^{-1}(\Delta; \Delta') = \frac{2}{(2\beta + \gamma)(1 - a)}
\begin{cases}
\Delta^{1-a}(\Delta')^{a-\xi} & \text{for} \quad \Delta \le \Delta', \\
(\Delta')^{1-\xi} & \text{for} \quad \Delta \ge \Delta',
\end{cases}
$$

is the Green function of $-M$ with the absorbing boundary condition (2.66) at $\Delta = 0$. It is again built by gluing two zero modes of the operator M, a constant and Δ^{1-a}. Now, the second one is more regular around $\Delta = 0$ and the first one around $\Delta = \infty$. The above formula for the stationary 2-point structure function stays valid in the intermediate compressibility phase for the choice of the absorbing boundary condition (physically, for very small viscosity and vanishing diffusivity) [EVa00].

Finally, for completeness, let us cite the result about the stationary 2-point structure function in the intermediate phase with the sticky choice (2.68) of boundary conditions (i.e. for fine-tuned small diffusivity and small viscosity):

$$
S^{(2)}(\Delta) = \frac{4}{2\beta + \gamma}\left(\int_0^\Delta (\Delta')^{-a}\, d\Delta' \int_0^{\Delta'} (\Delta'')^{a-\xi}\chi(\Delta'')\, d\Delta'' \right. \\
\left. + \frac{\mu}{1-a}\Delta^{1-a}\chi(0) \right),
$$

see [Gaw04].

2.9.5 Problems to Lecture 5

Problem 5.1 It follows from eqs. (2.61) and (2.62) that for $r \neq 0$,

$$\lim_{\substack{\xi \to 0 \\ \wp = \text{const.}}} \left(D^{ij}(0) - D^{ij}(r) \right) = \tfrac{1}{2} \gamma \delta^{ij}$$

as β has to tend to zero in this limit. Recall that the operator $\mathcal{M}_N$ takes the form (2.77) in the action on translation-invariant functions. Ignoring the contributions from $r_m = r_n$ with $m \neq n$ (they may be omitted in the weakly compressible regime), we obtain from that relation:

$$\lim_{\substack{\xi \to 0 \\ \wp = \text{const.}}} \mathcal{M}_N f(\underline{r}) = -\frac{\gamma}{4} \sum_{m \neq n} \nabla_{r_m^i} \nabla_{r_n^j} f(\underline{r}) = \frac{\gamma}{4} \sum_{n=1}^{N} \nabla_{r_n}^2 f(\underline{r}),$$

where we have used again the translation-invariance of f to get the last equality.

Problem 5.2 This is straightforward. Using the last expression for $\mathcal{M}_N$ and the formulae

$$\nabla_r^2 |r - r'|^2 = 2d, \qquad \nabla_r^2 |r - r'|^4 = 4(d+2)|r - r'|^2,$$

we obtain, denoting $r_i - r_j \equiv r_{ij}$, the relations

$$\mathcal{M}_4 \, |r_{12}|^2 |r_{34}|^2 = \gamma d \left(|r_{12}|^2 + |r_{34}|^2 \right),$$
$$\mathcal{M}_4 \left(|r_{12}|^4 + |r_{34}|^4 \right) = 2\gamma(d+2) \left(|r_{12}|^2 + |r_{34}|^2 \right)$$

from which it follows that f_0^4 is a zero mode. In general, translation-invariant zero modes of $\mathcal{M}_N$ at $\xi = 0$ are harmonic polynomials annihilated by the (Nd)-dimensional Laplacian.

Problem 5.3 Using the PDF (2.80), we obtain

$$S^{(N)}(\Delta) = \int \vartheta^N P_\Delta(\vartheta) \, d\vartheta = e^{BN + \frac{1}{2}AN^2}$$
$$= \left(\frac{\Delta}{\Delta_0} \right)^{\left(\frac{2-\xi}{2} + \frac{(1+2\wp)\xi}{d+2} \right) N - \frac{(1+2\wp)\xi}{2(d+2)} N^2}$$

for even N. This result does not mean, however, that for small ξ or $\frac{1}{d}$, the PDF of the scalar difference over small distance Δ becomes log-normal because, in reality, the perturbative scaling of higher powers sets in for smaller ξ or $\frac{1}{d}$. For fixed ξ and d, one observes numerically a quick saturation of the scalar scaling exponents for growing N to N-independent values. Such saturation signals abundant presence of fronts in the scalar field realizations [Cel01b].

Bibliography

[Adz01] L. Ts. Adzhemyan, N. V. Antonov, V. A. Barinov, Yu. S. Kabrits, A. N. Vasil'ev, "Calculation of the anomalous exponents in the rapid-change model of passive scalar advection to order ε^3", Phys. Rev. **E 64** (2001), 056306/1-28

[Adz02] L. Ts. Adzhemyan, N. V. Antonov, A. N. Vasil'ev, "Renormalization group, operator product expansion and anomalous scaling in models of passive turbulent advection", Acta Physica Slovaca **52** (2002), 541-546

[Ang06] L. Angheluta, R. Benzi, L. Biferale, I. Procaccia, F. Toschi, "Anomalous scaling exponents in nonlinear models of turbulence", Phys. Rev. Lett. **97** (2006), 160601/1-4

[Ant84] R. A. Antonia, E. Hopfinger, Y. Gagne, F. Anselmet, "Temperature structure functions in turbulent shear flows", Phys. Rev. **A 30** (1984), 2704-2707

[Ara01] I. Arad, L. Biferale, A. Celani, I. Procaccia and M. Vergassola, "Statistical conservation laws in turbulent transport", Phys. Rev. Lett. **87** (2001), 164502/1-4

[Arn03] L. Arnold, *Random Dynamical Systems*, Springer, Berlin 2003

[Bal01] E. Balkovsky, G. Falkovich, A. Fouxon, "Clustering of inertial particles in turbulent flows", chao-dyn/9912027 and "Intermittent distribution of inertial particles in turbulent flows", Phys. Rev. Lett. **86** (2001), 2790-2793

[Bal99] E. Balkovsky, A. Fouxon, "Universal long-time properties of Lagrangian statistics in the Batchelor regime and their application to the passive scalar problem", Phys.Rev. E, **60**, (1999) 4164-4174

[Bal00] E. Balkovsky and A. Fouxon and V. Lebedev, "On the Turbulent Dynamics of Polymer Solutions", Phys. Rev. Lett. **84** (2000), 4765-4768

[Ban06] M. M. Bandi, J. R. Cressman Jr., W. I. Goldburg, "Test of the Fluctuation Relation in compressible turbulence on a free surface", arXiv:CD/0607037

[Bat67] G. K. Batchelor, *An Introduction to Fluid Dynamics*, Cambridge University Press 1967

[Bax88] P. H. Baxendale, D. W. Stroock, "Large deviations and stochastic flows of diffeomorphisms", Prob. Theor.& Rel. Fields **80** (1988), 169-215

[Bec05] J. Bec, "Multifractal concentrations of inertial particles in smooth random flows", J. Fluid Mech. **528** (2005), 255-277

[Bec07] J. Bec, L. Biferale, M. Cencini, A. Lanotte, S. Musacchio, F. Toschi, "Heavy particle concentration in turbulence at dissipative and inertial scales", Phys. Rev. Lett. **98** (2007), 084502/1-4

[Bec06] J. Bec, M. Cencini, R. Hillerbrand, "Clustering of heavy particles in the

104

inertial range of turbulence" arXiv:nlin.CD/0606038

[Bec04] J. Bec, K. Gawędzki, P. Horvai, "Multifractal clustering in compressible flows", Phys. Rev. Lett. **92** (2004), 224501-2240504

[Ben06] R. Benzi, B. Levant, I. Procaccia, E. S. Titi, "Statistical properties of nonlinear shell models of turbulence from linear advection models: rigorous results", arXiv:nlin.CD/0612033

[Ber96] D. Bernard, K. Gawędzki, A. Kupiainen, "Anomalous scaling in the N-point functions of a passive scalar", Phys. Rev. **E 54** (1996), 2564-2572

[Ber98] D. Bernard, K. Gawędzki, A. Kupiainen, "Slow modes in passive advection", J. Stat. Phys. **90** (1998), 519-569

[Bif05] L. Biferale, G. Boffetta, A. Celani, B. J. Devenish, A. Lanotte, F. Toschi, "Lagrangian statistics of particle pairs in homogeneous isotropic turbulence", Phys. Fluids, **17** (2005), 115101/1-9

[Bir87] R. B. Bird, C. F. Curtiss, R. C. Armstrong, O. Hassager, *Dynamics of Polymeric Liquids, Vol. 2, Kinetic Theory*, Wiley, New York 1987

[Bof02] G. Boffeta, M. Cencini, M. Falcioni, A. Vulpiani, "Predictability: a way to characterize complexity", Phys. Rep. **356** (2002), 367-474

[Boh98] T. Bohr, M. H. Jensen, G. Paladin, A. Vulpiani, *Dynamical Systems Approach to Turbulence*, Cambridge University Press 1998

[Bof06] G. Boffetta, J. Davoudi, F. De Lillo, "Multifractal clustering of passive tracers on a surface flow", Europhys. Lett., **74** (2006), 62-68

[Bon06] F. Bonetto, G. Gallavotti, G. Gentile, "A fluctuation theorem in a random environment", mp_arc/06-139

[Bre68] L. Breiman, *Probability*, Addison-Wesley, Reading MA 1968

[Cel01a] A. Celani, M. Vergassola, "Statistical Geometry in Scalar Turbulence", Phys. Rev. Lett. **86** (2001), 424-427

[Cel01b] A. Celani, A. Lanotte, A. Mazzino, M. Vergassola, "Fronts in passive scalar turbulence", Phys. Fluids **13** (2001), 1768-1783

[Cha03] M. Chaves, K. Gawędzki, P. Horvai, A. Kupiainen, M. Vergassola "Lagrangian dispersion in Gaussian self-similar ensembles" J. Stat. Phys. **113** (2003), 643-692

[Che00] M. Chertkov, "Polymer Stretching by Turbulence", Phys. Rev. Lett. **84** (2000), 4761-4764

[Che96] M. Chertkov, G. Falkovich, "Anomalous scaling exponents of a white-advected passive scalar", Phys. Rev. Lett. **76** (1996), 2706-2709

[Che95] M. Chertkov, G. Falkovich, I. Kolokolov, V. Lebedev, "Normal and anomalous scaling of the fourth-order correlation function of a randomly advected scalar", Phys. Rev. **E 52** (1995), 4924-4941

[Che98] M. Chertkov, I. Kolokolov, M. Vergassola, "Inverse versus direct cascades in turbulent advection", Phys. Rev. Lett. **80** (1998), 512-515

[Che03] M. Chertkov and V. Lebedev, "Boundary effects on chaotic advection-diffusion chemical reactions", Phys. Rev. Lett. **90** (2003), 134501/1-4

[Che06] R. Chetrite, J.-Y. Delannoy, K. Gawędzki, "Kraichnan flow in a square: an example of integrable chaos", arXiv:nlin.CD/0606015, J. Stat. Phys. in press

[EVa00] W. E, E. Vanden Eijnden, "Generalized flows, intrinsic stochasticity, and turbulent transport", Proc. Nat. Acad. Sci. **97** (2000), 8200-8205

[EVa01] W. E, E. Vanden Eijnden, "Turbulent Prandtl number effect on passive scalar advection", Physica D **152-153** (2001), 636-645

[Eck81] J.-P. Eckmann, "Roads to turbulence in dissipative dynamical systems" Rev. Mod. Phys. **53** (1981), 643 - 654

[Eva93] D. J. Evans, E. G. D. Cohen, and G. P. Morriss, "Probability of second law violations in shearing steady states", Phys. Rev. Lett. **71** (1993), 2401-2404 and 3616

[Eva94] D. J. Evans, D. J. Searles, "Equilibrium microstates which generate the second law violating steady states" Phys. Rev. **E 50** (1994), 1645-1648

[Fal01] G. Falkovich, K. Gawedzki, M. Vergassola, "Particles and fields in fluid turbulence", Rev. Mod. Phys. **73** (2001), 913-975

[Fri99] U. Frisch, A. Mazzino, A. Noullez, M. Vergassola, "Lagrangian method for multiple correlations in passive scalar advection", Phys. Fluids **11** (1999), 2178-2186

[Gal95] G. Gallavotti, E. D. G. Cohen, "Dynamical ensembles in non-equilibrium statistical mechanics", Phys. Rev. Lett. **74** (1995), 2694-2697

[Gaw04] K. Gawędzki, P. Horvai, "Sticky behavior of fluid particles in the compressible Kraichnan model", J. Stat. Phys. **116** (2004), 1247-1300

[Gaw95] K. Gawędzki, A. Kupiainen, "Anomalous scaling of the passive scalar", Phys. Rev. Lett. **75** (1995), 3834-3837

[Gaw00] K. Gawędzki, M. Vergassola, "Phase transition in the passive scalar advection", Physica **D 138** (2000), 63-90

[Gra88] P. Grassberger, R. Baddi, A. Politi, "Scaling laws for invariant measures on hyperbolic and nonhyperbolic attractors", J. Stat. Phys. **51** (1988), 135-178

[Gro01] A. Groisman and V. Steinberg, "Efficient mixing of liquids at low Reynolds numbers using polymer additives", Nature **410** (2001), 905-908

[Hor05] P. Horvai, "Lyapunov exponent for inertial particles in the 2D Kraichnan model as a problem of Anderson localization with complex valued potential", arXiv:nlin.CD/0511023

[Kol41] A. N. Kolmogorov, "The local structure of turbulence in incompressible viscous fluid for very large Reynolds' numbers", C. R. Acad. Sci. URSS **30** (1941), 301-305

[Kra68] R. H. Kraichnan, "Small-scale structure of a scalar field convected by turbulence", Phys. Fluids **11** (1968), 945-963

[LeJ02] Y. Le Jan, O. Raimond, "Integration of Brownian vector fields", Ann. Probab. **30** (2002), 826-873

[LeJ04] Y. Le Jan, O. Raimond, "Flows, coalescence and noise", Ann. Probab. **32** (2004), 1247-1315

[Maj99] A. J. Majda and P. R. Kramer, "Simplified models for turbulent diffusion: Theory, numerical modelling and physical phenomena", Physics Reports **314** (1999), 237-574

[Meh05] B. Mehlig, M. Wilkinson, K. Duncan, T. Weber, M. Ljunggren, "On the aggregation of inertial particles in random flows" Phys. Rev. **E 72** (2005), 051104/1-10

[Moi01] F. Moisy, H. Willaime, J. S. Andersen, P. Tabeling, "Passive Scalar Intermittency in Low Temperature Helium Flows", Phys. Rev. Lett. **86** (2001), 4827 - 4830

[Oks03] B. Oksendal, *Stochastic Differential Equations*, 6th ed., Universitext, Springer, Berlin 2003

[Ols83] M. A. Olshanetsky, A. M. Perelomov: "Quantum integrable systems related to Lie algebras" Phys. Rep. **94** (1983), 313-404

[Ose68] V. I. Oseledec, "Multiplicative ergodic theorem: characteristic Lyapunov exponents of dynamical systems", Trudy Moskov. Mat. Obšč. **19** (1968), 179-210

[Pav06] G. A. Pavliotis, A. M. Stuart, "An introduction to multiscale methods", lecture notes, http://www.maths.warwick.ac.uk/ stuart/mult.html

[Ric26] L. F. Richardson, "Atmospheric diffusion shown on a distance-neighbour graph", Proc. R. Soc. Lond. **A 110** (1926), 709-737

[Ris89] H. Risken, *The Fokker Planck equation*, Springer, Berlin, 1989

[Rue79] D. Ruelle, "Ergodic theory of differentiable dynamical systems", Publications Mathématiques de l'IHÉS **50** (1979), 275-320

[Rue96] D. Ruelle, "Positivity of entropy production in nonequilibrium statistical mechanics", J. Stat. Phys. **85** (1996), 1-23

[Rue97] D. Ruelle, "Positivity of entropy production in the presence of a random thermostat", J. Stat. Phys. **86** (1997), 935-951

[Shr95] B. I. Shraiman, E. D. Siggia, "Anomalous scaling of a passive scalar in turbulent flow", C.R. Acad. Sci.**321** (1995), 279-284

[Sok99] I. M. Sokolov, "Two-particle dispersion by correlated random velocity fields", Phys. Rev. **E 60** (1999), 5528-5532

[Tay21] G. I. Taylor, "Diffusion by continuous movements", Proc. London Math. Soc. **20** (1921), 196-212

[Tel05] T. Tél, A. de Moura, C. Grebogi, G. Károlyi, "Chemical and biological activity in open flows: A dynamical system approach", Phys. Rep. **413** (2005), 91-196

[Tsi04] B. Tsirelson, "Nonclassical stochastic flows and continuous products", Probab. Surveys **1** (2004), 173-298

[You02] L.-S. Young, "What are SRB measures, and which dynamical systems have them" J. Stat. Phys. **108** (2002), 733-754

3

John Cardy. Reaction-diffusion processes

3.1 Introduction

The aim of this course is to introduce reaction-diffusion systems. These are non-equilibrium systems of diffusing classical particles, which undergo reactions such as pairwise annihilation. The system is governed by a master equation, but this may be expressed equivalently as a many-body quantum Hamiltonian. This allows perturbative solutions to correlation functions, including the mean number of particles in the system and density-density correlation functions. We will see that these systems, although non-equilibrium systems, have much in common with equilibrium systems governed by Langevin equations.

3.2 Brownian motion

As our first example of an equilibrium system, let us consider Brownian motion, which describes the motion of a mesoscopic particle, such as a grain of pollen, immersed in a bath of much smaller particles. We will work in one dimension for simplicity and set the mass of the particle to 1. Newton's equation is

$$\dot{v} = -\Gamma v + \xi(t) + F_{\text{ext}}, \tag{3.1}$$

in which v is the velocity of the particle, Γ is the strength of the Stokes friction, ξ is a random stochastic term drawn from a distribution with zero mean and F_{ext} is an external force. This stochastic equation is an example of a Langevin equation. We will examine more general Langevin type equations later. Note that the friction term may also be written as the derivative of the Hamiltonian, H.

$$-\Gamma v = -\Gamma \frac{\partial}{\partial v}(H = v^2/2). \tag{3.2}$$

The stochastic noise is characterised by its expectation values

$$\langle \xi(t) \rangle = 0 \tag{3.3}$$

$$\langle \xi(t)\xi(t') \rangle = f(t - t') \sim e^{-|t-t'|/\tau}, \tag{3.4}$$

with τ the typical collision time, which is assumed short compared to all other time scales in the problem. A small particle may collide with the pollen, but will then have its velocity randomised by interactions with other small particles. Hence, later collisions are essentially uncorrelated. Since we are interested in the behaviour on much longer time scales, we will take a delta function form for the second moment of the noise

$$\langle \xi(t)\xi(t') \rangle = 2D\delta(t - t'). \tag{3.5}$$

Note that $\xi(t)$ is not a differentiable function, so the machinery of stochastic calculus shoulde be used to interpret equation (3.1), but as we will always be integrating ξ over small times, we will sidestep this technicality.

3.2.1 The Einstein relation

Let us take $F_{ext} = 0$, then the system will come into thermal equilibrium with the heat bath at temperature T. A result of the equipartition theorem is $\langle v^2 \rangle = k_B T$, which may be used to derive a relation between Γ and D. Starting from equation 3.1,

$$v(t + \delta t) = (1 - \Gamma \delta t)v(t) + \int_t^{t+\delta t} \xi(t')dt' + O(\delta t^2) \tag{3.6}$$

$$\langle v(t + \delta t)^2 \rangle = (1 - 2\Gamma \delta t)\langle v(t)^2 \rangle + 2 \int_t^{t+\delta t} \langle v(t)\xi(t') \rangle dt'$$

$$+ \int_t^{t+\delta t} \int_t^{t+\delta t} \langle \xi(t')\xi(t'') \rangle dt'dt'' + O(\delta t^2). \tag{3.7}$$

The expectation value $\langle v(t)\xi(t') \rangle$ vanishes by causality, since the noise should be independent of the velocity at an earlier time. We may then substitute in equation (3.5) and use the equilibrium condition $\langle v(t+\delta t)^2 \rangle = \langle v(t)^2 \rangle = k_B T$ to find

$$k_B T = (1 - 2\Gamma \delta t)k_B T + 2D\delta t + O(\delta t^2). \tag{3.8}$$

Equating terms of order δt yields the Einstein relation for systems in equilibrium,

$$\Gamma k_B T = D. \tag{3.9}$$

3.2.2 Correlation function

We may also calculate the velocity two-point correlation function from equation 3.1. Let us continue to take $F_{\text{ext}} = 0$. The Fourier transform is

$$-i\omega\tilde{v}(\omega) = -\Gamma\tilde{v}(\omega) + \tilde{\xi}(\omega). \tag{3.10}$$

Hence

$$\langle\tilde{v}(\omega)\tilde{v}(\omega')\rangle = \frac{1}{-i\omega + \Gamma}\frac{1}{-i\omega' + \Gamma}\langle\tilde{\xi}(\omega)\tilde{\xi}(\omega')\rangle. \tag{3.11}$$

The Fourier transform of equation 3.5 is

$$\langle\tilde{\xi}(\omega)\tilde{\xi}(\omega')\rangle = 2D\delta(\omega + \omega'). \tag{3.12}$$

Substituting this result into equation 3.11

$$\langle\tilde{v}(\omega)\tilde{v}(\omega')\rangle = \frac{1}{\omega^2 + \Gamma^2}\delta(\omega + \omega'). \tag{3.13}$$

We may Fourier transform back to time

$$\langle v(t)v(t')\rangle = \int_{-\infty}^{\infty}\frac{d\omega}{2\pi}\frac{2De^{-i\omega(t-t')}}{\omega^2 + \Gamma^2} = k_BTe^{-\Gamma|t-t'|}, \tag{3.14}$$

which shows that the velocity fluctuations are correlated over the relaxation time $1/\Gamma$.

3.2.3 Response function

Let us add a term $-f(t)v(t)$ to the energy, corresponding to a linear coupling to the variable $v(t)$ and examine how this affects the expectation value of $v(t)$. In ω space, equation 3.1 now takes the form

$$-i\omega\langle\tilde{v}(\omega)\rangle = -\Gamma\langle\tilde{v}(\omega)\rangle + \Gamma\tilde{f}(\omega). \tag{3.15}$$

Solving for $\langle\tilde{v}(\omega)\rangle$,

$$\langle\tilde{v}(\omega)\rangle = \frac{f(\omega)}{-i\omega/\Gamma + 1} = G(\omega)f(\omega). \tag{3.16}$$

This defines the response function $G(\omega)$. In t space this is a convolution

$$\langle v(t)\rangle = \int_{-\infty}^{\infty}G(t - t')f(t')dt'. \tag{3.17}$$

Hence this is equivalent to the definition in terms of the functional derivative of $\langle v(t)\rangle$.

$$G(t - t') = \frac{\delta\langle v(t)\rangle}{\delta f(t')}\bigg|_{f=0}. \tag{3.18}$$

For Brownian motion, therefore, the correlation and response functions are related by

$$C(\omega) = \frac{2k_B T}{\omega} \operatorname{Im}[G(\omega)].$$ (3.19)

This is an example of the fluctuation-dissipation relation (FDT). We will derive the FDT for a more general Langevin equation in the next section.

3.3 More general Langevin equations

We have examined Brownian motion as an example of a stochastic process governed by a Langevin equation. We will now derive the correlation and response functions for a more general stochastic process with a single degree of freedom, labelled $\phi(t)$. The Langevin equation takes the form

$$\frac{d}{dt}\phi(t) = -\Gamma \frac{\partial H(\phi(t))}{\partial \phi(t)} + \xi(t),$$ (3.20)

where $\xi(t)$ is again a random term with zero time average and two-point correlation function

$$\langle \xi(t)\xi(t') \rangle = 2D\delta(t - t').$$ (3.21)

Equation 3.20 is equivalent to

$$\phi(t + \delta t) = \phi(t) - \Gamma \frac{\partial H(\phi(t))}{\partial \phi(t)} \delta t + \int_t^{t+\delta t} \xi(t')dt'.$$ (3.22)

Assuming that the system will relax to equilibrium in the presence of the noise, take the square of the above equation and average with respect to the Gibbs measure $e^{-H(\phi)/k_B T}$ to find

$$\langle \phi(t + \delta t)^2 \rangle - \langle \phi(t)^2 \rangle = -2\Gamma \langle \phi(t) \frac{dH(\phi(t))}{d\phi(t)} \rangle \delta t + 2D\delta t + O(\delta t^2).$$ (3.23)

The term $\langle \phi(t)H'(\phi(t)) \rangle$ may be integrated by parts and shown to be equal to $k_B T$, so equating terms of order δt, again leads to the Einstein relation

$$D = \Gamma k_B T.$$ (3.24)

3.3.1 The response function formalism

Expectation values of quantities such as $\phi(t_1)\phi(t_2)$ may be formally evaluated using a functional integral

$$\langle \phi(t_1)\phi(t_2) \rangle = \langle \int \phi(t_1)\phi(t_2)\delta[\phi(t) = \text{solution}]D\phi(t) \rangle_{\xi(t)},$$ (3.25)

where the delta function ensures that $\phi(t)$ is solution to the Langevin equation 3.20 and the average on the right hand side is now with respect to realisations of the noise $\xi(t)$. The delta function may be rewritten as

$$\delta[\phi(t) = \text{solution}] = \delta[\frac{d}{dt}\phi(t) + \Gamma\frac{\partial H(\phi(t))}{\partial \phi(t)} - \xi(t)] \times \text{Jacobian}, \qquad (3.26)$$

with the Jacobian equal to unity, as equation 3.20 is being interpreted as an Itô stochastic equation, the Jacobian is the determinant of an upper triangular matrix with 1 on the diagonal. The delta function may be expressed as a functional integral.

$$\delta[\frac{d}{dt}\phi + \Gamma\frac{dH(\phi)}{d\phi} - \xi] = \int e^{-\int \tilde{\phi}(t)[\frac{d}{dt}\phi(t)+\Gamma H'(\phi(t))-\xi(t)]dt} D\tilde{\phi}(t). \qquad (3.27)$$

The integral is over the field $\tilde{\phi}(t)$, which is known as the response field. In this form, the average over realisations of the noise may be taken using

$$\langle e^{\int \tilde{\phi}(t)\xi(t)dt}\rangle_\xi = e^{\frac{1}{2}\int\int dt'dt''\tilde{\phi}(t')\tilde{\phi}(t'')\langle\xi(t')\xi(t'')\rangle} = e^{D\int \tilde{\phi}(t)^2 dt}. \qquad (3.28)$$

Putting all this together, correlation functions, such as those in equation 3.25, may be expressed as

$$\langle\phi(t_1)\phi(t_2)\rangle = \int\int \phi(t_1)\phi(t_2)e^{-S[\tilde{\phi}(t),\phi(t)]} D\phi D\tilde{\phi}, \qquad (3.29)$$

where the effective action, S, is

$$S = \int \left[\tilde{\phi}(t)[\frac{d}{dt}\phi(t) + \Gamma H'(\phi(t))] - D\tilde{\phi}(t)^2\right]dt. \qquad (3.30)$$

This is the response function formalism. Response functions may be obtained in this formalism by adding a linear coupling to ϕ in the Hamiltonian, $H \to H - f(t)\phi(t)$. Then $S \to S - \Gamma \int f(t)\tilde{\phi}(t)dt$ and

$$\langle\phi(t_1)\rangle = \int\int e^{-S[\phi(t),\tilde{\phi}(t)]+\Gamma\int f(t)\tilde{\phi}(t)dt} D\phi(t)D\tilde{\phi}(t). \qquad (3.31)$$

From the derivative of this equation follows the response function.

$$G(t_2 - t_1) = \frac{\partial\langle\phi(t_1)\rangle}{\partial f(t_2)}\Big|_{f=0}$$

$$= \Gamma \int\int \phi(t_1)\tilde{\phi}(t_2)e^{-S[\phi(t),\tilde{\phi}(t)]} D\phi(t)D\tilde{\phi}(t). \qquad (3.32)$$

Alternatively completing the square in $\tilde{\phi}(t)$ in equation 3.31 and shifting integration variables $\tilde{\phi}(t) \rightarrow \tilde{\phi}(t) + \Gamma f(t)/2D$, the response function is

$$G(t_1 - t_2) = \frac{-\Gamma}{2D}\langle \phi(t_1)[\frac{d}{dt}\phi(t_2) + \Gamma H'(\phi(t_2))]\rangle$$

$$= \frac{\Gamma}{2D}\dot{C}(t_1 - t_2) - \frac{\Gamma^2}{2D}\langle \phi(t_1)H'(\phi(t_2))\rangle. \tag{3.33}$$

Let us consider the symmetries of the above equation under time reversal. The response function is zero for $t_1 < t_2$ by causality. The derivative of the correlation function is odd with respect to time. The term involving the Hamiltonian is even under exchange of the times t_1 and t_2, since the equilibrium average is time reversal invariant. We therefore conclude that for $t_1 > t_2$

$$G(t_1 - t_2) = \frac{\Gamma}{D}\dot{C}(t_1 - t_2) = k_B T \dot{C}(t_1 - t_2). \tag{3.34}$$

This is the fluctuation-dissipation relation again. In Fourier space, the FDT takes the form

$$\tilde{C}(\omega) = \frac{2k_B T}{\omega}\text{Im}[\tilde{G}(\omega)]. \tag{3.35}$$

3.3.2 The master equation

The master equation is a first order linear equation expressing the rate at which a system moves between states labelled by $\{\alpha\}$. At time t, let the system be in state α with probability $P(\alpha, t)$ and consider the time derivative of $P(\alpha, t)$. The change in P is due to transitions into and out of the state α. We will denote the rates of these processes by $R_{\alpha \rightarrow \beta}$. Then the master equation is

$$\frac{dP(\alpha, t)}{dt} = \sum_\beta R_{\beta \rightarrow \alpha} P(\beta, t) - R_{\alpha \rightarrow \beta} P(\alpha, t)$$

$$\equiv -\sum_\beta H_{\alpha\beta} P(\beta, t). \tag{3.36}$$

The conservation of probability requires that

$$0 = \frac{d}{dt}\sum_\alpha P(\alpha) = \sum_{\beta,\alpha} R_{\beta \rightarrow \alpha} P(\beta, t) - R_{\alpha \rightarrow \beta} P(\alpha, t). \tag{3.37}$$

In terms of $H_{\alpha\beta}$, this is

$$\sum_\alpha H_{\alpha\beta} = 0, \tag{3.38}$$

so H has a zero eigenvalue.

3.3.3 Detailed balance

The equilibrium distribution $P(\alpha) \propto e^{-H(\alpha)/k_B T}$ is stationary so $\partial_t P = 0$. Therefore,

$$\sum_\beta R_{\beta \to \alpha} P(\beta, t) - R_{\alpha \to \beta} P(\alpha, t) = 0. \tag{3.39}$$

The system satisfies detailed balance if this relation is valid term by term, ie

$$\frac{R_{\beta \to \alpha}}{R_{\alpha \to \beta}} = \frac{e^{-H(\alpha)/k_B T}}{e^{-H(\beta)/k_B T}}. \tag{3.40}$$

This is the case for the Metropolis algorithm used in numerical simulations of equilibrium systems. For this example,

$$\tilde{H}_{\alpha\beta} = e^{+H(\alpha)/k_B T} H_{\alpha\beta} e^{-H(\beta)/k_B T} \tag{3.41}$$

is symmetric ($\tilde{H}_{\alpha\beta} = \tilde{H}_{\beta\alpha}$), so has real eigenvalues. For the non-equilibrium particle systems we shall study, detailed balance is not valid.

3.4 Stochastic particle systems

We shall now turn our attention to non-equilibrium systems, in particular reaction-diffusion problems. These are classical systems with particles localised in space. These particles may be molecules, biological entities, fluctuating commodities in a market etc. There may be several species of particles in a given model, which will be labelled A, B, C and which reside on a lattice Z^d, labelled by site labels $\{j\}$. They undergo diffusion with characteristic diffusion constants D_A, D_B and react with rates λ, μ when inhabit the same lattice site. An example of such a reaction is represented by the equation

$$A + B \to C. \tag{3.42}$$

The first example we shall consider contains a single particle species, A, undergoing two particle annihilation, $A + A \to 0$. The steady state is not interesting; it is a single particle or zero particles, depending on whether the initial state of the system has an odd or an even number of particles. The interest in this model is the approach to the steady state. The starting point is the master equation. The states of the system are defined by the number of particles at each lattice site $\{n_j\}$ and the rates in the master equation are related to the reaction rate λ.

Although the system is a classical system, it may be re-expressed as a many-body quantum problem. The machinery associated with quantum mechanics then allows perturbative solutions to the correlation and response functions. The Fock space is composed of a vacuum state with zero particles, along with linear particle creation operators at each site $a_j^\dagger$. A general state is

$$|\{n_j\}\rangle = \prod_j a_j^{\dagger\,n_j}|0\rangle. \tag{3.43}$$

There are also particle annihilation operators, which are the Hermitian conjugates of the creation operators and remove particles at site j. The algebra satisfied by the operators is

$$[a_j, a_k^\dagger] = \delta_{j,k} \tag{3.44}$$

$$[a_j^\dagger, a_k^\dagger] = [a_j, a_k] = 0. \tag{3.45}$$

Using this, it may be shown that the states are eigenstates of the particle number operators $\{n_j\} = \{a_j^\dagger a_j\}$ with eigenvalue equal to the number of particles at site j. We can associate a state in the Fock space with a set of probabilities at time t

$$|\psi(t)\rangle = \sum_{\{n_j\}=(0,0\ldots)}^{\infty} p(\{n_j\}, t) \prod_j a_j^{\dagger\,n_j}|0\rangle. \tag{3.46}$$

With this definition, the master equation may be rewritten as a Schrodinger-type equation

$$\frac{\mathrm{d}}{\mathrm{d}t}|\psi(t)\rangle = -H(\{a\}, \{a^\dagger\})|\psi(t)\rangle. \tag{3.47}$$

Note that there are some differences from many-body quantum mechanics. The Schrodinger equation is real, so this is like Euclidean quantum mechanics. Our Hamiltonian is not (necessarily) hermitian. It may be shown that Hamiltonians coming from systems which satisfy detailed balance may be made symmetric and real by a similarity transformation. Also, the states are linear functions of the probabilities, rather than linear functions of the probability amplitudes as in quantum mechanics. We will return to this when we consider expectation values of observables, but first let us look at some examples of master equations and derive the associated Hamiltonians.

3.4.1 Particles hopping on a lattice

Consider a lattice consisting of two sites, 1 and 2. Particles hop from site 1 to site 2 only, with a rate D. The master equation is

$$\frac{dP(n_1, n_2, t)}{dt} = D(n_1 + 1)P(n_1 + 1, n_2 - 1, t) - Dn_1 P(n_1, n_2, t). \quad (3.48)$$

We may multiply the above equation by $a_1^{\dagger\, n_1} a_2^{\dagger\, n_2}|0\rangle$ and sum over all values of n_1, n_2. With the definition of the state $|\psi(t)\rangle$

$$|\psi(t)\rangle = \sum_{n_1, n_2} p(n_1, n_2, t) a_1^{\dagger\, n_1} a_2^{\dagger\, n_2}|0\rangle, \quad (3.49)$$

we may rewrite the equation obtained as

$$\frac{d|\psi(t)\rangle}{dt} = D \sum_{n_1, n_2} \left[p(n_1 + 1, n_2 - 1, t)(n_1 + 1) - p(n_1, n_2)n_1 \right] a_1^{\dagger\, n_1} a_2^{\dagger\, n_2}|0\rangle. \quad (3.50)$$

Subsituting the ladder operators for the number operators and using their commutation relations, we obtain

$$\frac{d|\psi(t)\rangle}{dt} = D \sum_{n_1, n_2} p(n_1 + 1, n_2 - 1, t) a_2^{\dagger} a_1 a_1^{\dagger\, n_1 + 1} a_2^{\dagger\, n_2 - 1}|0\rangle$$

$$- D \sum_{n_1, n_2} p(n_1, n_2, t) a_1^{\dagger} a_1 a_1^{\dagger\, n_1} a_2^{\dagger\, n_2}|0\rangle. \quad (3.51)$$

With a relabelling of indices in the first sum we have arrived at the desired formula, of the form of equation 3.47, with

$$H = -D(a_2^{\dagger} - a_1^{\dagger})a_1. \quad (3.52)$$

Had we also allowed hopping from site 2 to site 1 at the same rate, we would have obtained

$$H = D(a_2^{\dagger} - a_1^{\dagger})(a_2 - a_1). \quad (3.53)$$

3.4.2 Two particle annihilation

As a second example, consider two particle annihilation $A + A \to 0$ on a single lattice site. The master equation is

$$\frac{dP(n)}{dt} = \lambda(n + 2)(n + 1)P(n + 2) - \lambda n(n - 1)P(n). \quad (3.54)$$

In terms of $|\psi(t)\rangle$, this is

$$\frac{d|\psi(t)\rangle}{dt} = \lambda \sum_n (n+2)(n+1)P(n+2,t)a^{\dagger n}|0\rangle - \lambda \sum_n n(n-1)P(n,t)a^{\dagger n}|0\rangle$$

$$= \lambda \sum_n P(n+2)a^2 a^{\dagger n+2}|0\rangle - \lambda \sum_n P(n)a^{\dagger 2}a^2 a^{\dagger n}|0\rangle . \qquad (3.55)$$

Hence, we obtain a Schrodinger-type equation with

$$H = -\lambda(1 - a^{\dagger 2})a^2 . \qquad (3.56)$$

Combining the two equations above, we find the Hamiltonian for the lattice annihilation model

$$H = D \sum_{\langle ij \rangle} (a_i^\dagger - a_j^\dagger)(a_i - a_j) - \lambda \sum_i (1 - a_i^{\dagger 2})a_i^2 , \qquad (3.57)$$

where the first sum is over pairs of nearest neighbour sites.

3.4.3 Averages of observables in the many-body formalism

For a given master equation, we have shown how to derive a Schrodinger-like equation for the evolution of the state of the system in the form of equation 3.47. This may be integrated as usual to find

$$|\psi(t)\rangle = e^{-Ht}|\psi(0)\rangle . \qquad (3.58)$$

A convenient choice for the initial probability distribution is an independent Poisson distribution at each site

$$P(\{n_i\}, t = 0) = \prod_j e^{-\rho_0} \frac{\rho_0^{n_j}}{n_j!} , \qquad (3.59)$$

since this corresponds to the initial state of the system being a coherent state

$$|\psi(0)\rangle = \prod_j e^{-\rho_0} e^{\rho_0 a_j^\dagger}|0\rangle . \qquad (3.60)$$

The average value of an observable $A(\{n_j\})$ is defined as

$$\overline{A} = \sum_{\{n_j\}} A(\{n_j\})p(\{n_j\}, t) . \qquad (3.61)$$

Using the identity $\langle 0|e^a a^{\dagger n}|0\rangle = 1$, this may be rewritten as

$$\overline{A} = \langle 0| \prod_j e^{a_j} A(\{n_j\})|\psi(t)\rangle . \qquad (3.62)$$

As an example, let us calculate the expectation value of the identity operator:

$$1 = \langle 0| \prod_j e^{a_j} e^{-H(\{a_j, a_j^\dagger\})t} |\psi(0)\rangle. \tag{3.63}$$

This can only be satisfied for all t if $\langle 0| \prod_j e^{a_j} H(\{a_j, a_j^\dagger\}) = 0$. Using $\langle 0|e^a a^\dagger = \langle 0|e^a$, we see that the Hamiltonian must satisfy

$$H(\{a_j\}, \{a_j^\dagger = 1\}) = 0. \tag{3.64}$$

This is an expression of the conservation of probability.

3.4.4 The Doi shift

The term $\prod_j e^{a_j}$ may be commuted through to the right in equation 3.62, by making use of the identity

$$e^a f(a^\dagger) = f(a^\dagger + 1)e^a. \tag{3.65}$$

Note that this is simply a consequence of the commutation relations for the ladder operators. The resulting form of the expectation value of observables is

$$\langle 0|A(\{a_j^\dagger + 1\}, \{a_j\})e^{-H(\{a_j^\dagger + 1\}, \{a_j\})t} e^{\sum_j a_j} |\psi(0)\rangle. \tag{3.66}$$

With the choice of a coherent state for the initial wavefunction, the ket takes the simple form

$$e^{\sum_j a_j} |\psi(0)\rangle = e^{\sum_j \rho_0 a_j^\dagger} |0\rangle. \tag{3.67}$$

The Hamiltonian with $\{a_j^\dagger\} \to \{a_j^\dagger + 1\}$ is known as the Doi shifted Hamiltonian. For our example of particles hopping on the lattice and pairwise annihilating, the shifted Hamiltonian takes the form

$$H_{\text{shifted}} = D \sum_{\langle ij \rangle} (a_i^\dagger - a_j^\dagger)(a_i - a_j) + \lambda \sum_j 2a_j^\dagger a_j^2 + a_j^{\dagger 2} a_j^2. \tag{3.68}$$

Note that this Hamiltonian is normal ordered, with the consequence that its vacuum expectation value is zero.

3.4.5 Path integral representation

From the many-body description, we may use perturbation theory to calculate expectation values of observables. First it is convenient to re-express expectation values in terms of path integrals. Split a given time interval t into small increments δt and recall the identity

$$e^{-Ht} = \lim_{\delta t \to 0} (1 - \delta t H)^{t/\delta t} . \tag{3.69}$$

A complete set of coherent states

$$1 = \int \frac{\mathrm{d}\phi \mathrm{d}\phi^*}{\pi} e^{-\phi^*\phi} e^{\phi a^\dagger} |0\rangle \langle 0| e^{\phi^* a} , \tag{3.70}$$

may be inserted into equation 3.62 at each time step. Thus, the expectation value is a product of terms like

$$\langle 0| e^{\phi^*(t+\delta t)a} [1 - \delta t H(a^\dagger, a)] e^{\phi(t)a^\dagger} |0\rangle . \tag{3.71}$$

The exponentials may be commuted to the other side of the square bracket using equation 3.65 and the conjugate expression to obtain

$$\langle 0| 1 - \delta t H(a^\dagger + \phi^*, a + \phi)|0\rangle e^{\phi^*(t+\delta t)\phi(t)} . \tag{3.72}$$

As the Hamiltonian is normal ordered, this is equivalent to

$$\langle 0| 1 - \delta t H(\phi^*, \phi)|0\rangle e^{\phi^*(t+\delta t)\phi(t)} . \tag{3.73}$$

Re-exponentiating the product of terms involving the Hamiltonian and taking the continuum time limit, we see that averages of observables are integrals over a pair of fields on the discrete lattice, with an effective action

$$\prod_j \left[\int \int \frac{\mathrm{d}\phi_j(t)\mathrm{d}\phi_j^*(t)}{\pi} \right] e^{-S[\phi^*,\phi]} \tag{3.74}$$

$$S[\phi^*, \phi] = \int \mathrm{d}t \sum_j \phi_j^* \partial_t \phi_j + H(\{\phi_j^*\}, \{\phi_j\}) , \tag{3.75}$$

where the time dependence of ϕ and ϕ^* has been omitted. The action for $A + A \to 0$ is

$$S = \int \mathrm{d}t \sum_j \phi_j^* \partial_t \phi_j + D \sum_{\langle ij \rangle} (\phi_i^* - \phi_j^*)(\phi_i - \phi_j) + \lambda \sum_j (2\phi_j^* \phi_j^2 + \phi_j^{*2}\phi_j^2) . \tag{3.76}$$

Just as in the Hamiltonian, $a^\dagger$ operators in the observables are replaced by ϕ^* fields and a operators by ϕ. We can now take a naive continuum limit, converting the product of integrals into functional integrals over the fields $\phi(t)$ and $\phi^*(t)$ and recasting S for our example in the form

$$S = \int \mathrm{d}t \int \mathrm{d}^d x \left[\phi^* \partial_t \phi + D\nabla \phi^* \nabla \phi + \lambda(2\phi^* \phi^2 + \phi^{*2}\phi^2) \right] . \tag{3.77}$$

Originally, ϕ^* was taken to be the complex conjugate of ϕ, but we may now treat them as independent variables, relabelling ϕ^* as $\tilde\phi$. After an integration by parts on the gradient terms, we obtain for our Lagrangian

$$\tilde\phi[\partial_t \phi - D\nabla^2 \phi + 2\lambda\tilde\phi\phi^2] + \lambda\tilde\phi^2\phi^2 . \tag{3.78}$$

This is the same form as the Langrangian we would obtain from an equilibrium problem with the Langevin equation

$$\partial_t \phi = D\nabla^2\phi - 2\lambda\phi^2 + \xi . \tag{3.79}$$

In the next subsection we will show that $\langle \phi(t) \rangle = \bar{n}(t)$, the expected number of particles at time t. Taking expectation values of the above equation therefore yields

$$\frac{dn}{dt} = D\nabla^2 n - 2\lambda n^{(2)} , \tag{3.80}$$

where $n^{(2)} = \langle \phi(t)^2 \rangle$ is the probability of finding 2 particles at the same site at time t. If we approximate $n^{(2)} = n^2$, we obtain the rate equation. Since a diffusion process starting from a Poisson distribution in the absence of noise remains a Poisson distribution at later times, we deduce that the noise describes deviations from a Poissonian distribution. This noise is not white noise, however. Recall from section 3.3.1 that starting from white noise with the correlation function $\langle \xi(x,t)\xi(x',t') \rangle = 2D\delta(x'-x)\delta(t'-t)$, we obtain a term in the Lagrangian of the form $-D\tilde\phi^2$. Here, however, the Lagrangian contains a term $\lambda\tilde\phi^2$, so the correlations of the noise are

$$\langle \xi(x,t)\xi(x',t') \rangle = -2\lambda\phi^2\delta(x'-x)\delta(t'-t) . \tag{3.81}$$

Hence, the noise is complex. If we start with a real field ϕ, it too becomes complex. There is a physical reason why we cannot simply have real white noise in this problem. A given particle, which hasn't annihilated, will have swept out an area around it without any particles contained inside. So, particles must be anti-correlated. For the connected part of $\langle \phi(x_1,t_1)\phi(x_2,t_2) \rangle$ to be negative, we require negatively correlated noise since the response function is positive

$$\langle \phi(x_1,t_1)\phi(x_2,t_2) \rangle_c =$$
$$\int\int G(t_1-t',x_1-x')G(t_2-t'',x_2-x'')\langle \xi(x',t')\xi(x'',t'') \rangle dt'dx'dt''dx'' . \tag{3.82}$$

The explanation for this result is that ϕ, as a fluctuating quantity is not the same as the density. It is the expectation values of the two which are equal.

One of the problems is to show that

$$\bar{n}^2 = \langle \phi^2 \rangle + \langle \phi \rangle \,, \tag{3.83}$$

from which it is clear that $n \neq \phi$. The vanishing of $\langle \phi^2 \rangle - \langle \phi \rangle^2$ would simply result in $\bar{n}$ having a Poissonian distribution.

3.4.6 The expected number of particles and the expectation value of ϕ

In this short subsection, we will show that, on average, ϕ is the same as n, the number of particles.

$$\bar{n} = \langle 0 | e^a a^\dagger a e^{-Ht} | \psi(0) \rangle$$
$$= \langle 0 | e^a a e^{-Ht} | \psi(0) \rangle, . \tag{3.84}$$

In the path integral picture

$$\langle 0 | e^a a e^{-Ht} | \psi(0) \rangle = \frac{\int \int D\phi D\tilde{\phi} \phi e^{-S}}{\int \int D\phi D\tilde{\phi} e^{-S}} = \langle \phi \rangle \,. \tag{3.85}$$

So $\bar{n} = \langle \phi \rangle$.

3.5 Feynman diagrams and the renormalization group

The form of the Lagrangian allows us to write down the propagator and vertex diagrams and so to formulate pertubative solutions to the correlation functions of the theory. We will not proceed in this way, but instead start from the stochastic equation

$$\partial_t \phi(x,t) = D\nabla^2 \phi(x,t) - 2\lambda\phi(x,t)^2 + \xi(x,t) + \rho_o \delta(t) \,, \tag{3.86}$$

with the following correlation function for the noise

$$\langle \xi(x,t)\xi(x',t') \rangle = -2\lambda\phi(x,t)^2 \delta(t-t')\delta(x-x') \,. \tag{3.87}$$

Note that the term $\rho_o \delta(t)$ comes from the Doi shifted initial Poissonian (coherent) state. We find the formal solution to this differential equation

$$\phi(x,t) = \int G_0(x-x',t-t')[-2\lambda\phi(x',t') + \xi(x',t') + \rho_0 \delta(t)] \,, \tag{3.88}$$

where the Green's function, $G_0(x,t)$, satisfies the equation

$$(\partial_t - D\nabla^2)G_0(x,t) = \delta(x)\delta(t) \,. \tag{3.89}$$

Equation 3.88 permits an iterative solution, which is most easily described pictorially in terms of the Feynman diagrams in Fig 3.1. Time increases from right to left, starting from $t = 0$ at the right hand side.

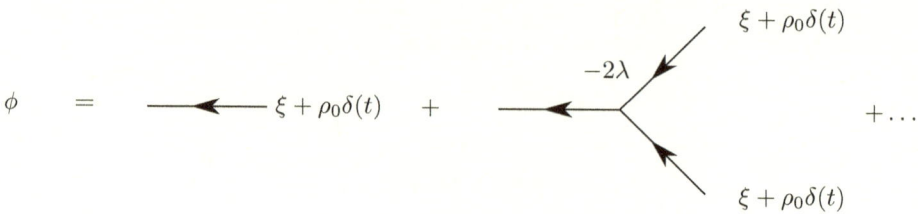

Fig. 3.1. Feynman diagrams contributing to ϕ.

So, ϕ is seen to be a sum of tree diagrams. If we switch off the noise, $\xi \to 0$, we obtain the diagrams in Fig 3.2.

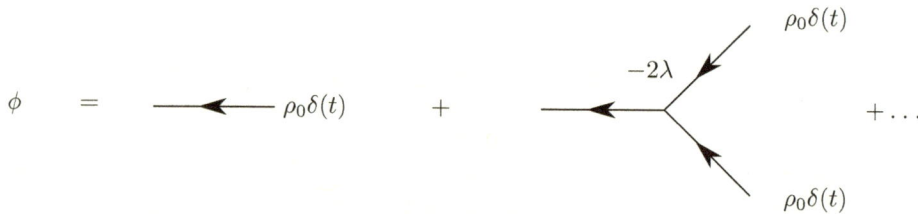

Fig. 3.2. Feynman diagrams contributing to ϕ in the absence of noise.

These diagrams may be expressed alternatively as a recursive relation, described pictorially in Fig 3.3. Algebraically, these diagrams represent

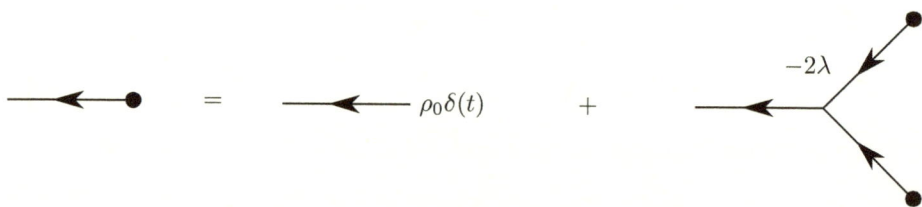

Fig. 3.3. The recursive definition of ϕ.

$$\phi(t) = \int d^d x \, G_0(t, x) \rho_0 - 2\lambda G_0 \circ \phi^2 \,, \tag{3.90}$$

where $\circ$ denotes a convolution. This can be readily converted back to a differential equation

$$\partial_t \phi = D\nabla^2 \phi - 2\lambda \phi^2 + \rho_0 \delta(t) \,, \tag{3.91}$$

which is just the rate equation and shows that our results are consistent. If we assume spatial homogeneity, the rate equation has a simple solution

$$\phi = \frac{\rho_0}{1 + 2\rho_0 \lambda t} ,$$ (3.92)

which has the long time behaviour

$$\phi(t \to \infty) = \frac{1}{2\lambda t} .$$ (3.93)

Note that this long time solution is independent of ρ_0. Now let us reinstate the noise and examine the average of ϕ, using

$$\langle \xi(x,t) \rangle = 0$$ (3.94)

$$\langle \xi(x,t)\xi(x',t') \rangle = -2\lambda \phi(t)^2 \delta(x - x')\delta(t - t')$$ (3.95)

$$\langle \xi^4 \rangle = \sum_{\text{pairs}} \langle \xi^2 \rangle \langle \xi^2 \rangle .$$ (3.96)

The last identity follows from ξ having a Gaussian distribution. It is equivalent to Wick's theorem in quantum field theory. Averaging leads to the pairwise contraction of ξ insertions, leading to diagrams with loops like those in Fig 3.4.

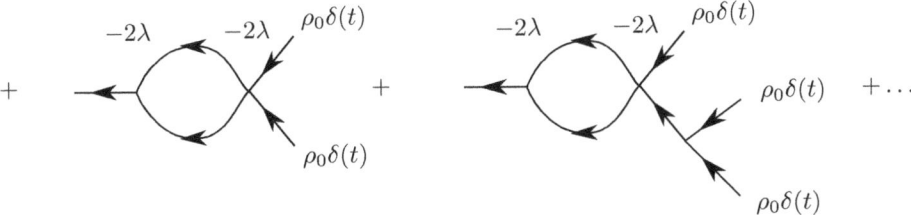

Fig. 3.4. The average over noise of ϕ.

Each diagram is constructed from the set of components in Fig 3.5. Those familiar with quantum field theory would have been able to identify these directly from the action. The set of all diagrams may be decomposed into the

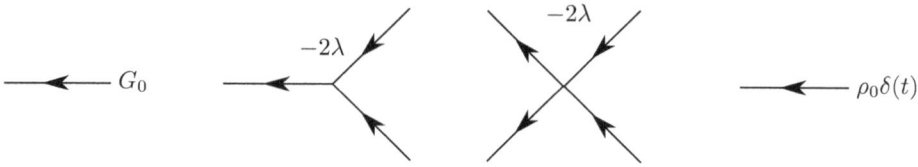

Fig. 3.5. The propagators and vertices in $\langle\phi\rangle$.

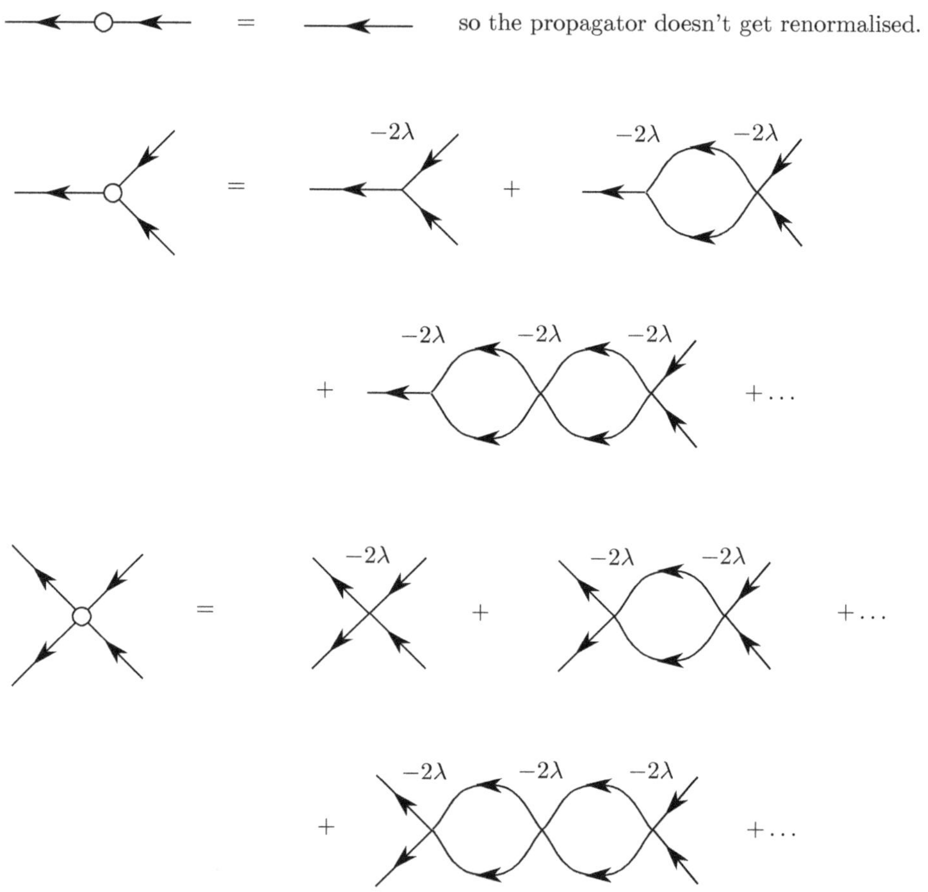

so the propagator doesn't get renormalised.

Fig. 3.6. The full propagator and vertex diagrams

skeleton diagrams shown in Fig 3.6. The loops do not affect the propagators, but lead to corrections to the vertices. In fact, both vertices are renormalized in the same way. This is a consequence of probability conservation ensuring that there is only one coupling constant, λ. To calculate the Feynman diagrams, it is easier to consider the Green's functions in Fourier space.

$$G_0(x,t) = \int_0^\infty \frac{ds}{2\pi i} \int_{-\infty}^\infty \frac{d^d\mathbf{k}}{(2\pi)^d} \frac{e^{st+i\mathbf{k}\cdot\mathbf{x}}}{s + Dk^2} . \tag{3.97}$$

In order to perform the vertex renormalizations, it is necessary to calculate the loop diagram in Fig 3.7:

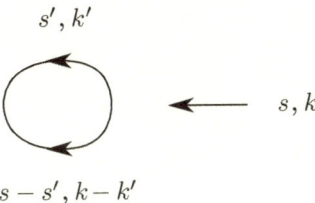

Fig. 3.7. One loop diagram

$$\int_0^\infty \frac{ds'}{2\pi i} \int_{-\infty}^\infty \frac{d^d k'}{(2\pi)^d} \frac{1}{s' + Dk'^2} \frac{1}{s - s' + D(k - k')^2} . \tag{3.98}$$

Integrating over s' and taking $k = 0$

$$= \int_{-\infty}^\infty \frac{d^d k'}{(2\pi)^d} \frac{1}{s + 2Dk'^2} . \tag{3.99}$$

The integral appears to diverge at large k' for $d > 2$, but this is a consequence of discarding the lattice cut-off, which restricts the upper limit on k'. The divergence at $s = 0$ for $d < 2$ is more interesting. Above two dimensions, the integral is an analytic function of s. For $d \geq 2$, the loop corrections to the density give non-leading terms and simply renormalize ρ_0 and λ. We see then that the rate equation is (qualitatively) asymptotically correct for $d \geq 2$. The loop diagram may be thought of as a correction due to annihilations between particles which have survived occupying the same site at an earlier time without annihilating. The dimensionality of the system is important because it determines the dimensionality of the random walk in relative particle coordinates. For $d = 1$, a random walk returns with almost certain probability to a point in space in finite time. For $d > 2$, the random walk almost certainly does not return in finite time, so corrections corresponding to the loop diagram above are irrelevant. $d_c = 2$ is known as the critical dimension of the theory. The renormalization group is a way of resumming series which are divergent in the limit $s \to 0$, such as those with loops in figure 3.6. Define $\lambda_R(s)$ as the sum of diagrams contributing to the full three point propagator in figure 3.6 with total outgoing frequency s. We see that the diagrams are convolutions in (t, x) space, so are geometric series in (s, q) space. Using equation 3.99 and doing the integral over k':

$$\lambda_R(s) = \frac{\lambda_0}{1 + \frac{k_d}{\epsilon} \frac{\lambda_0 s^{-\epsilon/2}}{D^{1-\epsilon/2}}}, \tag{3.100}$$

where $\epsilon = 2 - d$ and k_d is finite for $d = 2$. It is defined as

$$k_d = \frac{2\epsilon}{(8\pi)^{d/2}} \Gamma(1 - d/2). \tag{3.101}$$

It is easiest to work with a dimensionless coupling constant. Looking at the denominator of equation 3.100, a suitable choice is

$$g_R(s) = \lambda_R s^{-\epsilon/2} D^{\epsilon/2 - 1}. \tag{3.102}$$

The expected number of particles may be expressed as a function of the bare coupling constant, or alternatively as a function of the renormalized coupling constant and the scale s. This scale was introduced arbitrarily, though, so at fixed λ_0 the expected number of particles should not depend on s.

$$s \frac{d}{ds}\bigg|_{\lambda_0} \overline{n}(t, g_R, s) = 0, \tag{3.103}$$

where we have assumed that the asymptotic form of $\overline{n}$ does not depend on the initial density n_0. This assumption is not necessary for this derivation, but shortens the calculation. g_R is a function of s, so this may be rewritten as

$$(s \frac{\partial}{\partial s} + \beta(g_R) \frac{\partial}{\partial g_R}) \overline{n}(t, g_R, s) = 0, \tag{3.104}$$

where the derivative with respect to s acts only on the explicit s dependence of $\overline{n}$ and the beta function is defined by

$$\beta(g_R) \equiv s \frac{\partial}{\partial s} g_R(s)\bigg|_{\lambda}. \tag{3.105}$$

For our example, from equation 3.102

$$\beta(g_R) = \frac{-\epsilon}{2} g_R + \frac{k_d}{2} g_R^2. \tag{3.106}$$

Dimensional analysis determines the form of $\overline{n}$

$$\overline{n}(t, g_R, s) \sim (t)^{-d/2} f(st, g_R), \tag{3.107}$$

where f is some function of its dimensionless parameters. This shows that

$$s\frac{\partial}{\partial s}\bar{n}(t, g_R, s) = \left(t\frac{\partial}{\partial t} + \frac{d}{2}\right)\bar{n}(t, g_R, s).$$ (3.108)

We have obtained the following partial differential equation for $\bar{n}$, known as the Callan-Symanzik equation:

$$\left(t\frac{\partial}{\partial t} + \beta(g_R)\frac{\partial}{\partial g_R} + \frac{d}{2}\right)\bar{n}(t, g_R, s) = 0.$$ (3.109)

The solution to this partial differential equation is obtained via the method of characteristics

$$\bar{n}(t, g_R, s) = (ts)^{-d/2}\bar{n}(t = s^{-1}, \tilde{g}(t), s),$$ (3.110)

where $\tilde{g}(t)$ is known as the running coupling and satisfies the equation

$$t\frac{\mathrm{d}}{\mathrm{d}t}\tilde{g}(t) = -\beta(\tilde{g}(t)).$$ (3.111)

For $d > 2$, ϵ is negative

$$t\frac{\mathrm{d}}{\mathrm{d}t}\tilde{g}(t) = -\left(-\frac{\epsilon}{2}\right)\tilde{g}(t),$$ (3.112)

so the running coupling flows to zero as $t \to \infty$. In this case, we are not justified in neglecting the long time dependence of $\bar{n}$ on n_0. For $d < 2$, $\tilde{g}(t)$ flows to the zero of the beta function $g^* = \epsilon/k_d$. We may evaluate $\bar{n}$ in the right hand side of equation 3.110 for small s, since this corresponds to late times. Therefore, we may use the long time solution of the rate equation 3.93 to find

$$\bar{n}(t) \sim (st)^{-d/2}\frac{1}{2\lambda s^{-1}}$$

$$\sim (t)^{-d/2}\frac{1}{2D^{1-\epsilon/2}g^*}$$

$$\sim (Dt)^{-d/2}A(\epsilon).$$ (3.113)

$A(\epsilon)$ is known as a universal amplitude, since it depends only on the dimensionality of the system. We have shown here that $A = (1/4\pi\epsilon)(1 + O(\epsilon))$. To find the higher order terms, we need to evaluate the higher loop diagrams in perturbation theory.

3.5.1 The critical dimension

At $d_c = 2$, the equation for the running coupling constant 3.111

$$t\frac{\mathrm{d}\tilde{g}(t)}{\mathrm{d}t} = \frac{-k_2}{2}\tilde{g}(t)^2$$ (3.114)

has solution

$$\tilde{g}(t) = \frac{4\pi}{\ln t}.$$ (3.115)

Hence, substituting this into the solution for $\bar{n}$, we obtain

$$\bar{n}(t) \sim \frac{1}{8\pi} \frac{\ln t}{Dt}.$$ (3.116)

The coefficient $1/8\pi$ is exact and universal and we see the well-known logarithmic corrections in the critical dimension.

3.6 Other reaction-diffusion processes

The methods we have learnt are applicable to a range of other reaction-diffusion processes. The general strategy is to write down a Hamiltonian, formulate a field theory and renormalize this field theory. We will examine two examples for which this method fails however, due to conservation laws. These laws lead to different types of fluctuations and slow dynamics.

3.6.1 $A + B \to 0$

The first example has two species, labelled A and B. For convenience, they will have identical properties, except their labels. They react via $A + B \to 0$ with rate λ. This reaction is realised in electrodynamics by the annihilation of oppositely charged particles. Note that the reaction conserves $n_a - n_b$ locally, since the reactions are local. From the master equation, the following Hamiltonian may be derived

$$H = H_{\text{hopping}} - \lambda \sum_j (a_j b_j - a_j^\dagger b_j^\dagger a_j b_j).$$ (3.117)

The path integral representation therefore has an effective action

$$S = \int dt \int d^d x \left[a^* \partial_t a + b^* \partial_t b + D\nabla a^* \nabla a + D\nabla b^* \nabla b - \lambda(ab - a^* b^* ab) \right].$$ (3.118)

S must be dimensionless, since it appears as the argument of an exponential function in the functional integral. In terms of units of wavenumber, k, the fields have the following dimensions

$$[aa^*] = [bb^*] = k^d$$ (3.119)

$$[\lambda] = 2 - d.$$ (3.120)

Hence, the upper critical dimension for the theory is $d_c = 2$. However, it turns out that in certain circumstances fluctuations are still important for

$d > 2$. We would like to investigate the evolution of the system from a spatially homogeneous initial state, such as when the distribution of A and B particles is an independent Poisson distribution at each site. For $d > 2$ we can use the inhomogeneous rate equations

$$\partial_t a = D\nabla^2 a - \lambda ab \qquad (3.121)$$

$$\partial_t b = D\nabla^2 b - \lambda ab, \qquad (3.122)$$

where $a = n_a$, etc. Subtract the above equations to see that $\langle \chi \rangle = \langle a - b \rangle$ satisfies a simple diffusion equation, with formal solution

$$\chi(x,t) = \int d^d x' G_0(t, x - x') \chi(x', 0). \qquad (3.123)$$

$\chi(x', 0)$ is a random variable with mean zero. However, $\chi(x, 0)$ is a fluctuating quantity. Assuming

$$\langle \chi(x', 0) \chi(x'', 0) \rangle = \Delta \delta(x' - x''), \qquad (3.124)$$

where $\Delta \propto \rho_0$, it follows that

$$\langle \chi(x,t)^2 \rangle = \Delta \int d^d x' G(t, x - x')^2 \sim \frac{\Delta}{t^{d/2}}. \qquad (3.125)$$

Define also $\phi \equiv a + b$, which satisfies the equation

$$\partial_t \phi = D\nabla^2 \phi - \lambda\phi^2 + \lambda\chi^2 + \text{noise}. \qquad (3.126)$$

Considering average values,

$$\partial_t \langle \phi \rangle = D\nabla^2 \langle \phi \rangle - \lambda \langle \phi^2 \rangle + \lambda \langle \chi^2 \rangle. \qquad (3.127)$$

For solutions with translational invariance, the ∇^2 term vanishes. We may replace $\langle \phi^2 \rangle$ with $\langle \phi \rangle^2$ to find an upper bound for our solution. Thus,

$$\partial_t \langle \phi \rangle = -\lambda \langle \phi \rangle^2 + \frac{\lambda \Delta}{t^{d/2}}. \qquad (3.128)$$

On setting $\Delta = 0$, we see that $\phi \sim 1/\lambda t$. This is asymptotically correct for $d > 4$ if $\Delta \neq 0$ by dimensional analysis. If $d < 4$, the two terms on the right hand side balance asymptotically, so that

$$\langle \phi \rangle \sim \sqrt{\frac{\Delta}{t^{d/2}}} \sim \frac{1}{t^{d/4}}. \qquad (3.129)$$

Hence, the critical dimension is 4 and so three dimensions displays interesting non-mean field behaviour. Locally $\phi^2 \approx \chi^2$, so $a + b = |a - b|$. This means that either $a \gg b$ or $a \ll b$. Therefore, most parts of the system are dominated by one particle type separated by relatively narrow reaction zones. This is known as segregation.

3.6.2 $A + A \rightleftharpoons C$

Consider now the process $A + A \rightleftharpoons C$, where the reaction proceeds from left to right at rate λ and from right to left at rate μ. The quantity $n_a + 2n_c$ is conserved by the reaction and this conservation law again leads to interesting dynamics. The rate equations are

$$\partial_t a = -2\lambda a^2 + 2\mu c$$
$$\partial_t b = \lambda a^2 - \mu c \,, \tag{3.130}$$

with the constraint that

$$\partial_t(a + 2c) = 0 \,. \tag{3.131}$$

The steady state solution is

$$\lambda a_\infty^2 = \mu c_\infty \,, \tag{3.132}$$

with

$$a_\infty + 2c_\infty = a_0 + 2c_0 \,. \tag{3.133}$$

Expanding the rate equations in terms of $c = c_\infty + \delta c$ and $a = a_\infty + \delta a$, we find an exponentially fast approach to the steady state $\delta a, \delta c \sim e^{-t/t_0}$, but this is not what actually happens; fluctuations play an important role. The Hamiltonian is

$$H = H_{\text{diff}} - \lambda(c^\dagger a^2 - a^{\dagger 2} a^2) - \mu(a^{\dagger 2} c - c^\dagger c) \,. \tag{3.134}$$

The Doi shifts $c^\dagger \to c^\dagger + 1$ and $a^\dagger \to a^\dagger + 1$ lead to the following form for the shifted Hamiltonian:

$$H_{\text{shifted}} = H_{\text{diff}} + \lambda(a^{\dagger 2} + 2a^\dagger - c^\dagger)a^2 - \mu(a^{\dagger 2} + 2a^\dagger - c^\dagger c) \,. \tag{3.135}$$

The effective Lagrangian is equivalent to the rate equations

$$\partial_t a = D\nabla^2 a - 2\lambda a^2 + 2\mu c + \xi \tag{3.136}$$
$$\partial_t b = D\nabla^2 b + \lambda a^2 - \mu c \,, \tag{3.137}$$

There is no noise term in equation 3.137, since the Lagrangian does not contain a $c^{\dagger 2}$ term. The noise in equation 3.136 satisfies

$$\langle \xi(x,t)\xi(x',t') \rangle = (\mu c - \lambda a^2)\delta(x - x')\delta(t - t') \,. \tag{3.138}$$

Note that the noise vanishes in the equilibrium steady state. This is not to say there are no fluctuations, but instead that the density distribution is a product of local Poisson distributions. The stochastic equation for $\chi = a + 2c$ is

$$\partial_t \chi = D\nabla^2 \chi + \xi \,. \tag{3.139}$$

This may be solved for a given realisation of the noise

$$\chi = \chi_0 + \int d^d x' \int dt\, G_0(x - x', t - t')\xi(x', t')\,. \tag{3.140}$$

χ_0 is the conserved piece so that $\langle \chi \rangle = \chi_0$. Let us look at the two point correlation function for χ

$$\langle \chi(t)^2 \rangle - \langle \chi(t) \rangle^2 = \int d^d x' \int dt'\, G_0(x - x', t - t')^2 \langle \mu c(x', t') - \lambda a^2(x', t') \rangle\,. \tag{3.141}$$

The expectation value is equivalent to $-\partial_{t'}\langle c(t') \rangle$ which is supposed to be independent of x'. Thus the spatial integral may be performed to find

$$\langle \chi(t)^2 \rangle - \langle \chi(t) \rangle^2 = -\int dt' \frac{1}{(t - t')^{d/2}} \frac{\partial}{\partial t'} \langle c(t') \rangle\,. \tag{3.142}$$

For large times, this integral is dominated by $t' \ll t$ and we obtain

$$\langle \chi(t)^2 \rangle - \langle \chi(t) \rangle^2 = \frac{c_0 - c_\infty}{t^{d/2}}\,. \tag{3.143}$$

We conclude that particles become correlated or anti-correlated depending on whether the initial density of C particles is larger than the equilibrium density or smaller than the equilibrium density. The asymptotic solution to the stochastic differential equations can be found systematically, but a faster route to the solution is to assume local equilibrium $\lambda a^2 = \mu c$. We also have that

$$a + 2c = a_0 + 2c_0 + \delta\chi\,. \tag{3.144}$$

Solving these equations locally gives

$$a = \frac{-\mu + \sqrt{\mu^2 + 8\mu\lambda(a_0 + 2c_0 + \delta\chi)}}{4\lambda}\,. \tag{3.145}$$

The fluctuations in $\delta\chi$ affect the approach to equilibrium of a since $a = f(\delta\chi)$ so that

$$\langle a \rangle = \langle f(\delta\chi) \rangle$$
$$= f(0) + \frac{1}{2}f''(0)\langle \delta\chi^2 \rangle\,. \tag{3.146}$$

As the fluctuations in $\delta\chi$ scale as $t^{-d/2}$, so does the approach to equilibrium of a

$$\langle a \rangle = a_\infty + Ct^{-d/2}\,, \tag{3.147}$$

where C is a calculable constant.

3.7 Reaction-diffusion systems and turbulence (*C. Connaughton, R. Rajesh and O. Zaboronski*)

The formalism of dynamical renormalization group introduced above in the context of reaction-diffusion (RD) systems may be applied to the study of turbulent systems. In this context, turbulence does not refer exclusively to hydrodynamic problems, but rather to a class of non-equilibrium systems having in common the following features.

The defining characteristic of a turbulent system is the existence of a stationary state with widely separated sources and sinks of some conserved quantity. These stationary states do not obey the detailed balance condition, instead they are characterised by a flux of a conserved quantity from the source to the sink. Universal statistics are expected in the inertial range, the region far from the source and sink. The best known example is the Navier–Stokes (N–S) turbulence at high Reynolds number with forcing at large scales [Frisch(1995)]. The steady state is characterised by an energy flux from large length scales to small length scales where the energy is dissipated by viscous forces. Other examples include Burgers turbulence [Falkovich et al.(2001)], the Kraichnan model of passive scalar advection [Falkovich et al.(2001)], kinematic magneto hydrodynamics [Falkovich et al.(2001)], and wave turbulence [Zakharov et al.(1992)]. A common feature of these systems is that they each admit a phenomenological description based on constancy of the flux and the assumption of self-similarity. The first such theory was the Kolmogorov 1941 (K41) theory of N–S turbulence. Obtaining a quantitative understanding of the limitations and applicability of such phenomenology is a major theoretical challenge. Analytic progress in hydrodynamics has been slow despite strong numerical and experimental evidence for many interesting and nontrivial violations of K41-style self-similarity. Studies of other systems, notably Burgers equation and the Kraichnan model, have been more successful [Falkovich et al.(2001)], providing insight into the breakdown of self-similarity.

There is a simple reaction-diffusion system, that is turbulent in the sense described above, the universal properties of which are described by a perturbative fixed point of dynamical renormalization group, [Connaughton et al.(2005)], [R. Rajesh et al.(2006)]. Namely, let us consider diffusing-coagulating massive point particles in a d-dimensional space. Assume that coagulations conserve mass. As the result of coagulations particles tend to grow heavier. By adding light particles to the system at a constant rate we can drive the system to a non-equilibrium steady state characterised by a constant flux of mass directed from small to large masses. It turns out that many properties of the steady state of this interacting particle system agree with the phenomenological picture of a turbulent steady state developed in the context of hydrodynamic turbulence.

3.7.1 Cluster-cluster aggregation: model and continuum description

Consider a d-dimensional lattice whose sites are occupied by particles indexed by a mass. At a given moment of time a configuration of the system is determined by specifying the set of occupation numbers, $\{N_t(\mathbf{x}_i, m)\}_{\mathbf{x}_i \in \mathbf{R}^d}^{m \in \mathbf{Z}^+}$, where $N_t(\mathbf{x}, m)$ is the number of particles of mass m on site $\mathbf{x}$ at time t. The configuration changes in time due to three processes.

Diffusion: A particle hops with rate D to one of its nearest neighbours. The diffusion constant D is taken to be independent of mass.

Aggregation: Two particles at the same lattice site aggregate at rate λ to form one particle whose mass is the sum of the constituent particles. The reaction rate λ is assumed to be independent of mass.

Injection: At each lattice site, particles of mass m_0 are injected at rate J/m_0.

We will call the above model the mass model (MM). The MM has three physical parameters: the diffusion constant D, the aggregation rate λ, and input mass flux J. In addition there are two parameters associated with the lattice - the lattice spacing, dx, and the smallest mass, m_0. Starting from the lattice model, the stochastic integro-differential equation describing the system may be obtained from the path integral formalism described in Sec. 3.4.5:

$$
\left(\frac{\partial}{\partial t} - D\nabla^2\right)\phi(m) = \lambda \int_0^m dm'\phi(m')\phi(m - m')
$$
$$
-2\lambda\phi(m)\int_0^\infty dm'\phi(m') + \frac{J}{m_0}\delta(m - m_0) + i\sqrt{2\lambda}\phi(m)\eta, \quad (3.148)
$$

where $\eta(\vec{x}, t)$ is white noise with $\langle\eta(\vec{x}, t)\eta(\vec{x}', t')\rangle = \delta(t - t')\delta^d(\vec{x} - \vec{x}')$ and $i = \sqrt{-1}$. As explained in Sec. 3.4.5, the imaginary multiplicative noise term accounts for inter-particle anti-correlations. All correlation functions of the mass distribution can be expressed in terms of the correlation functions of $\phi(m, \vec{x}, t)$. In particular,

$$
P_n(m) \sim \frac{1}{n!}\langle[\phi(m, \vec{x}, t)]^n\rangle\left[1 + O(m^{-1})\right], \quad (3.149)
$$

where $P_n(m_1, \ldots, m_n)(\Delta V)^n \prod_i dm_i$ is the probability of having particles with masses in the intervals $[m_i, m_i + dm_i]$ in a volume ΔV and $\langle \ldots \rangle$ denotes averaging with respect to the noise η. See problem (xii), section 3.8.1 for the derivation of this important result.

It is possible to map the MM to the $A + A \to A$ model reaction in the presence of a source. The reaction $A + A \to A$ is equivalent to the reaction $A + A \to \emptyset$ analysed in Section 3.4, see the solution to problem (vii) below. Therefore, the mapping of the aggregation problem to the single-species reaction will immediately enable the application of the renormalization group formalism developed above to aggregation. The reduction of (3.148) to the $A + A \to A$ Langevin equation is achieved by applying Laplace transform with respect to the mass variable [Zaboronski(2001)].† Let

$$R_\mu(\vec{x}, t) = \int_0^\infty dm \phi(\vec{x}, m, t) - \int_0^\infty dm \phi(\vec{x}, m, t) e^{-\mu m}. \tag{3.150}$$

Then,

$$\left(\frac{\partial}{\partial t} - D\nabla^2\right) R_\mu(\vec{x}, t) = -\lambda R_\mu^2 + \frac{j_\mu}{m_0} + 2i\sqrt{\lambda} R_\mu(\vec{x}, t)\eta(\vec{x}, t), \tag{3.151}$$

where $j_\mu = J(1 - e^{-\mu m_0})$. In terms of field $R_\mu(\vec{x}, t)$, eq. (3.151) becomes a stochastic version of the rate equation for the $A + A \to A$ reaction in the presence of a source. This immediately implies that the critical dimension of constant kernel aggregation model is $d_c = 2$. The computation of the average mass distribution in the MM reduces to solving a one-species particle system with a μ-dependent source and then computing the inverse Laplace transform with respect to μ. For example, to compute the average density, $\langle N_m(t) \rangle$, for the mass model, we first calculate $\langle R_\mu \rangle$, the average of the solution of eq. (3.151) with respect to the noise, $\eta(\vec{x}, t)$. We then take the inverse Laplace transform with respect to μ and obtain the density from eq. (3.150). To compute $P_n(m, t)$, correlation functions of the form $\langle R_{\mu_1}(\vec{x}, t) R_{\mu_2}(\vec{x}, t) \ldots R_{\mu_n}(\vec{x}, t) \rangle$ are needed. These are non-trivial, as the $R_\mu(\vec{x}, t)$'s are correlated for different μ's via the common noise term in Eq. (3.151), see Section 3.7.4 for more detials.

The Feynman rules and the propagator of the theory describing the perturbative solution of (3.151) for a fixed index μ are identical to that of $A + A \to A$ (see Sec. 3.5).

However, care must be taken to include the "extra" vertex which arises when computing correlations between fields with different μ indices. The

---

† Such a map exists for the constant aggregation kernel only. As a result, the constant kernel aggregation model is significantly simpler to analyse than a generic aggregation model, but it still retains many qualitative features of the generic case.

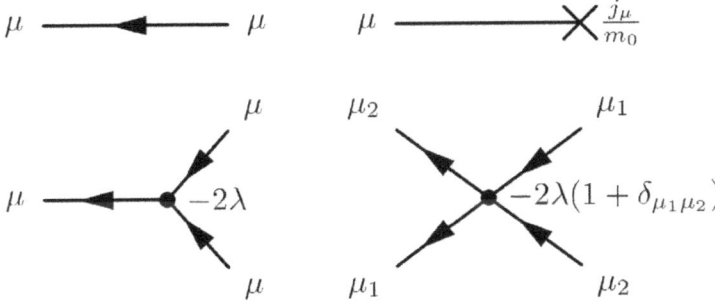

Fig. 3.8. Propagators and vertices of the theory.

Feynman rules are summarised in fig. 3.8 with time increasing from right to left. The slightly more complicated prefactor for the quartic vertex takes into account the aforementioned "extra" vertex.

3.7.2 Self-similar theory

We begin by asking what we can learn about possible stationary states of the MM from simple dimensional considerations. The first quantity of interest in characterising the long time behaviour of the model is the stationary mass spectrum, $\langle N_m \rangle$. $\langle N_m \rangle$ is the number of particles of mass m per unit volume in the stationary state. Specifically, we would like to know how $\langle N_m \rangle$ scales with m, when $m \gg m_0$. Since we are hoping for universal behaviour in the limit of large m, we assume that the mass spectrum does not depend on the position of the source, m_0. This assumption must be verified at a later stage. The spectrum does not depend on $\vec{x}$ due to translation invariance. The remaining dimensional parameters upon which $\langle N_m \rangle$ could, in principle, depend are J, D, λ and , of course, m. We shall perform most of our computations in units where $D = 1$ but for the purposes of the present analysis we will keep D dimensionful. The dimension of N_m is $[N_m] = M^{-1}L^{-d}$. The dimensions of the other parameters are: $[J] = ML^{-d}T^{-1}$, $[D] = L^2T^{-1}$ and $[\lambda] = L^dT^{-1}$. It is immediately evident that there are too many dimensional parameters in the model to uniquely determine the mass spectrum. One can readily verify that for any scaling exponent, x, and dimensionless constant, c_1, the formula

$$\langle N_m \rangle = c_1 \, J^{x-1} D^{\frac{(3-2x)d}{d-2}} \lambda^{\frac{(d+2)x-2d-2}{d-2}} m^{-x}, \qquad (3.152)$$

is a dimensionally correct expression for $\langle N_m \rangle$ which scales as m^{-x}. This is different to the dimensional argument used by Kolmogorov in his theory of hydrodynamic turbulence. For that system, there is a unique dimensionally correct combination of parameters giving the energy spectrum. Eq. (3.152) allows us to pick out the scaling exponent, x, for the reaction-limited and diffusion-limited regimes†:

$$x^{KZ} = \frac{3}{2}, \tag{3.153}$$

$$x^{K41} = \frac{2d+2}{d+2}. \tag{3.154}$$

The above two exponents correspond to a different balance of physical processes in order to realise a stationary state. We briefly discuss each to explain the choice of nomenclature.

We shall call x^{KZ} the Kolmogorov–Zakharov (K–Z) exponent since it is the analogue for aggregating particles of the K–Z spectrum of wave turbulence, see Section 1.2 of Falkovich' lectures and [Zakharov et al.(1992)], in the sense that it is obtained as the stationary solution of a mean field kinetic equation. This spectrum describes a reaction limited regime where diffusion plays no role.

We shall call x^{K41} the Kolmogorov 41 (K41) exponent since it is a closer analogue of the 5/3 spectrum of hydrodynamic turbulence originally proposed by Kolmogorov in his 1941 papers on self-similarity in turbulence. This is because in the Navier-Stokes equations there is no dimensional parameter like the reaction rate controlling the strength of the nonlinear interactions. This exponent describes a diffusion limited regime where the reaction rate, λ, plays no role, reactions being effectively instantaneous.

The case $x = 1$ corresponds to one in which $\langle N_m \rangle$ does not depend on the mass flux J. However, this is not of physical interest for this problem as it describes equipartition of mass incompatible with positive value of mass flux in the system. On the other hand, each of the regimes characterised by x^{KZ} and x^{K41} carry a mass flux and is relevant.

Note that the answer $x^{K41} = (2d+2)/(d+2)$ coincides with the result of an exact computation of the mass spectrum for $d < 2$ [Rajesh and Majumdar(2000)] so we expect that the Kolmogorov theory is correct in $d < 2$. In $d > 2$ it is known that $x = 3/2$. Therefore the KZ conjecture is appropriate in higher dimensions.

The success of scaling analysis for the mass model can be easily explained

---

† In the reaction-limited regime, diffusion is slow and reaction is fast, in diffusion-limited regime the opposite holds true.

using an RG argument. In the dimension $d > 2$, the field theory corresponding to (3.151) is non-renormalizable. As a result, the mass spectrum is given by the solution of mean field Smoluchowski equation which describes reaction-limited regime of the MM. In $d < 2$, we expect the reaction to be dominated by the fixed point of the RG flow for $A + A \rightarrow A$ reaction, see Sec.3.5. Therefore, the mass spectrum does not depend on the reaction rate. Furthermore, field Φ has no anomalous dimension, which justifies the naive dimensional argument, i. e. $K41$ conjecture.

We are interested in more than just the average mass density $\langle N_m \rangle$. To characterise correlations in the the MM we must also study the probability distributions of multi-particle configurations $P_n(m_1, \ldots, m_n)$ for $n > 1$. For instance, to analyse self-similarity of mass distribution, we should compute the homogeneity exponent, γ_n, of P_n defined through

$$P_n(\Gamma m_1, \ldots, \Gamma m_n) = \Gamma^{-\gamma_n} P_n(m_1, \ldots, m_n) \tag{3.155}$$

As before, let us start with dimensional analysis. The formula

$$P_n(m_1, \ldots, m_n)$$
$$= c_n \, J^{(\gamma_n - n)} D^{\frac{(3n - 2\gamma_n)d}{d-2}} \lambda^{\frac{(d+2)\gamma_n + (2d+2)n}{d-2}} (m_1 \ldots m_n)^{-\frac{\gamma_n}{n}}, \tag{3.156}$$

is dimensionally consistent for any exponent γ_n where c_n is a dimensionless constant. We are again assuming that the large mass behaviour of the P_n's is independent of m_0. The simplest way to obtain a theoretical prediction for the mass dependence of the P_n's is to use a self-similarity conjecture analogous to Kolmogorov's 1941 conjecture. Assume that P_n depends only on the masses m_i, mass flux J and either the reaction rate, λ or the diffusion coefficient D. For the reaction limited case we obtain

$$\gamma_n^{KZ} = \frac{3}{2} n, \tag{3.157}$$

and for the diffusion limited case,

$$\gamma_n^{K41} = \frac{2d + 2}{d + 2} n. \tag{3.158}$$

Note that in both cases, the dependence of γ_n on the index n is linear, reflecting the assumed self-similarity of the statistics of the local mass distribution.

The K41 self-similarity conjecture assumes that P_n does not depend on λ, m_0, the lattice spacing, and the box size $\Delta V dm_1 \ldots dm_n$. The lack of dependence on the lattice spacing is expected due to the renormalizability of the effective field theory describing the MM below two dimensions. We will however find an anomalous dependence of correlation functions on a length scale depending on the other parameters and the box size which leads to a violation of self-similarity.

We now show that the self-similar prediction for $P_1(m) = \langle N(m) \rangle$ is correct. Consider $\langle R_\mu \rangle$. Let R_{mf} be the sum of all tree diagrams with one outgoing line. R_{mf} satisfies the diagrammatic equation similar to the one shown in Fig. 3.3. It corresponds to the noiseless limit of Eq. (3.151). The solution is

$$R_{mf}(t) = \sqrt{\frac{j}{m_0\lambda}} \tanh\left(\sqrt{\frac{j\lambda}{m_0}}t\right) \overset{t\to\infty}{\longrightarrow} \sqrt{\frac{j}{m_0\lambda}}. \qquad (3.159)$$

It has been shown in Sec. 3.5 that the re-summation of the leading terms in the loop expansion renormalizes the coupling constant. (See also [Droz and Sasvári(1993)] for the corresponding computation for $A + A \to A$ reaction in the presence of input.) The renormalization law can be found exactly,

$$\lambda \to \lambda_R = C(\epsilon)L_D(1 + O(\mu)),$$

where $C(\epsilon)$ is a dimensionless constant. Replacing λ by λ_R in eq. (3.159) gives $\langle R_\mu \rangle$. Therefore $R_\mu \sim (J\mu)^{d/(d+2)}$ as $\mu \to 0$. The inverse Laplace transform gives

$$C_1(m) \sim (D^{-1}J)^{d/(d+2)}m^{-\gamma_1}, \quad d < 2, \qquad (3.160)$$

where $\gamma_1 = (2d+2)/(d+2)$. This agrees with the K41 prediction and the exact result. The K41 conjecture thus corresponds to renormalized mean field theory. A dimensional argument gives the correct scaling of R_μ since it has zero anomalous dimension. Hence $\langle R_\mu \rangle$ scales with its physical dimension. It turns out that this is simple scaling behaviour does not hold for multi-particle probabilities.

3.7.3 The conservation of mass and the counterpart of Kolmogorov 4/5-th law.

We know that the K41 hypothesis works for $n = 1$. It is possible to analytically confirm that $\gamma_n \neq \gamma_n^{K41}$ in $d < 2$ by calculating γ_2 exactly. From the definition of γ_2, it follows that

$$P_2(m_1, m_2) = \left(\frac{1}{m_1 m_2}\right)^{\gamma_2/2} \phi\left(\frac{m_1}{m_2}\right), \qquad (3.161)$$

where ϕ is an unknown scaling function which satisfies $\phi(x) = \phi(1/x)$ due to the permutation symmetry of P_2. Our aim is to compute γ_2 without using the ϵ-expansion introduced in Section 3.5.

In terms of the Laplace transform,

$$\Phi_2(m_1, m_2) = \int_{\sigma-i\infty}^{\sigma+i\infty} \int_{\sigma-i\infty}^{\sigma+i\infty} d\mu_1 d\mu_2 \langle R_{\mu_1} R_{\mu_2} \rangle e^{-m_1 \mu_1} e^{-m_2 \mu_2}, \quad (3.162)$$

where R_μ solves eq. (3.151). Due to eq. (3.161),

$$\langle R_{\mu_1} R_{\mu_2} \rangle = \left(\frac{1}{\mu_1 \mu_2} \right)^{1-\gamma_2/2} \psi \left(\frac{\mu_1}{\mu_2} \right), \quad (3.163)$$

where ψ is an unknown scaling function. To find the large m_1, m_2 asymptotics of Φ, we need to know the small μ_1, μ_2 asymptotics of $\langle R_{\mu_1} R_{\mu_2} \rangle$. Averaging eq. (3.151), with respect to noise and setting $\partial_t \langle R_\mu \rangle = 0$ in the large time limit, we find that $\langle R_\mu R_\mu \rangle = \frac{j}{\lambda m_0} \approx \frac{J\mu}{\lambda}$ for $\mu \ll m_0$. Comparing this result with eq. (3.163) we find that $\gamma_2 = 3$.

Note that γ_2 does not depend on dimension, d, of the lattice. Therefore, it is correctly predicted by mean field theory. The non-renormalization of γ_2 by diffusive fluctuations can be explained by mass conservation or, more precisely by constancy of the average flux of mass in the mass space, see [Connaughton et al.(2007)] for more details. Here, we simply wish to point out that the exact answers for γ_1 and γ_2 establish multi-scaling non-perturbatively: the points $(0,0)$, $(1,\gamma_1)$ and $(2,\gamma_2)$ do not lie on the same straight line.

Due to its close connection with mass conservation, the law $\gamma_2 = 3$ is a counterpart of the 4/5 law of Navier-Stokes turbulence. Recall, that 4/5 law states that the third order longitudinal structure function of the velocity field (the correlation function 'responsible' for energy flux) scales in the inertial range as the first power of the separation. It is interesting to notice, that Kolmogorov theory respects 4/5 law in Navier-Stokes turbulence, but violates $\gamma_2 = 3$ in the MM.

It is worth stressing that the result $\gamma_2 = 3$ reflecting the constancy of mass flux in the steady state is very robust, in the sense that it does not depend on the details of particle transport. Assume for example, that our diffusing-coagulating particles are also advected by a turbulent flow. As explained in the lecture courses by Falkovich and Gawedzki, the effect of advection can be modelled by adding the term

$$\vec{v}(\vec{x}, t) \cdot \nabla \phi(\vec{x}, t, m)$$

to the left hand side of the stochastic Smoluchowski equation (3.148). Here

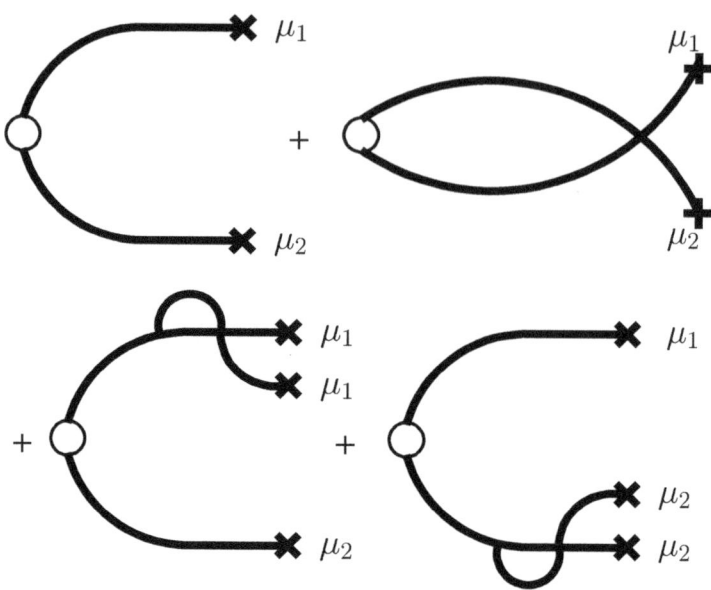

Fig. 3.9. Diagrams contributing to multi-point probabilities.

$\vec{v}(\vec{x}, t)$ is a turbulent velocity field. The expected value of this extra term is zero and the calculation of γ_2 explained above goes through yielding $\gamma_2 = 3$.

3.7.4 Higher order correlation functions

For $n > 2$ the details of the model are important and we need to employ the formalism of epsilon-expansion of Section 3.5 to compute γ_n.

Consider the diagrams contributing to $\langle R_{\mu_1} R_{\mu_2} \rangle$ up to one loop order shown in Fig. 3.9.

Diagrams 3 and 4 contribute to coupling constant renormalization of the mean field contribution of diagram 1. Diagram 2 is of a different nature: it generates the order ϵ contribution to the anomalous dimension of $\langle R_{\mu_1} R_{\mu_2} \rangle$. There are $\binom{n}{2}$ such diagrams contributing to the anomalous dimension of the n-point function. Calculating these diagrams gives the anomalous dimension of the n-point function as $-\epsilon n(n-1)/2$. The physical dimension is $-dn$. The scaling dimension is the sum of the two. Hence $\langle \prod_{i=1}^{n} R_{\mu_i} \rangle \sim \Phi_n \prod_{i=1}^{n} L_i^{-d-\epsilon(n-1)/2}$, where Φ_n is a scaling function of the variables L_i/L_j, where $L_i = [m_0 D/j(\mu_i)]^{1/(d+2)}$. Inverse Laplace transform

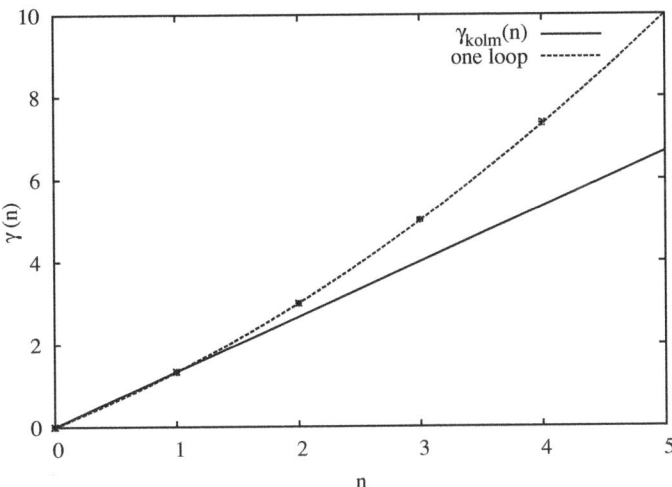

Fig. 3.10. Comparison of numerically measured γ_n in one dimension with the prediction of ϵ-expansion for $\epsilon = 1$.

gives

$$P_n(m_1, \ldots, m_n) \sim \Psi_n \prod_{i=1}^{n} \frac{1}{m_i} \left(\frac{J}{Dm_i} \right)^{\frac{d+\epsilon(n-1)/2}{d+2}} \tag{3.164}$$

Ψ_n is a scaling function of the variables m_i/m_j and the parameters ΔV, m_0 and λ. In the limit when $m_0 \to 0$ and $\lambda \to \infty$, we expect $\Psi_n \sim (\Delta V)^{\frac{\epsilon n(n-1)}{2d}}$. Note that $\Psi_n \to 0$ as $\Delta V \to 0$ which describes anti-correlation of particles. From Eq. (3.164),

$$\gamma_n = \left(\frac{2d+2}{d+2} \right) n + \left(\frac{\epsilon}{d+2} \right) \frac{n(n-1)}{2} + O(\epsilon^2). \tag{3.165}$$

The first term is γ_n^{SS} and the nonlinear terms correct SS scaling leading to a breakdown of self-similarity. The predictions up to $o(\epsilon)$ are also shown on figure 3.10. The closeness is surprising for a first order ϵ-expansion with $\epsilon = 1$. Note that the first order ϵ-expansion matches the exact answers for γ_1 and γ_2. We conjecture that the one-loop answer for γ_n is exact in one dimension.

The quadratic nature of the corrections in Eq. (3.165) is different from the analogous quadratic scaling exponents of velocity structure functions in the lognormal model of random energy cascade of N–S turbulence, see [Frisch(1995)] for a review. For large n, the lognormal model predicts negative scaling exponents which violate the Novikov inequality [Frisch(1995)]. In contrast Eq. (3.165) predicts that $\gamma(n)$ is greater than the SS value, reflecting the anti-correlation between particles.

In $d = 2$, nonlinear logarithmic corrections to SS are expected. These can be calculated exactly using the RG method. They vanish for $n = 2$ consistent with the exact result $P_2(m) \sim m^{-3}$. We present the final results here:

$$P_n(m) \sim \frac{J^{n/2}[\ln(m)]^{n-n^2/2}}{m^{3n/2}}\left[1+O\left(\frac{1}{\ln(m)}\right)\right], \quad d=2. \qquad (3.166)$$

3.7.5 Refined self similarity

We show that the results for MM is consistent with refined-self similarity conjecture due Kolmogorov and Obukhov (see [Frisch(1995)], page 164 for a review). The original conjecture states that the difference between the actual scaling exponent of the Navier-Stokes velocity structure function and the corresponding Kolmogorov value is equal to the scaling exponent of the energy flux fluctuation.

For the case of the mass aggregation model, the expression for the mass flux is

$$J(m) = \lambda \int dm \int dm' \int dm'' m\delta(m - m' - m'')N(m_1)N(m_2) + .. \quad (3.167)$$

where λ is the reaction rate. Therefore,

$$\langle J(m_1)J(m_2)\ldots J(m_n)\rangle \sim m^{3n}C_{2n}(m_1, m_2, \ldots, m_{2n}). \qquad (3.168)$$

If μ_n is the scaling exponent of $\langle J(m_1)J(m_2)\ldots J(m_n)\rangle$, than the above relationship reads

$$\gamma_n = \frac{3n}{2} - \mu_n/2. \qquad (3.169)$$

This is the exact counterpart of Kolmogorov-Obukhov's RSS conjecture: if the fluctuation of the mass flux around the average constant value is negligible, then $\mu_n = 0$ and $\gamma_n = 3n/2$ - its mean field value. Note in particular that $\mu_1 = 0$ as the average value of the flux is constant and does not depend on the measurement point in the mass space. Therefore, we recover $\gamma_2 = 3$ a consequence of constancy of mass flux. Another consequence of (3.169) is that strong fluctuations of the mass cascade always lead to the violation of Kolmogorov scaling. Combining (3.169) with (3.165) we find that in $d < 2$,

$$\mu_n = \frac{\epsilon}{2}n(n - 1) + O(\epsilon^2). \qquad (3.170)$$

3.8 Exercises

3.8.1 Problems

Several problems below use bosonic ladder operators. These are creation operators and annihilation operators which add and remove (respectively) particles from the system. The operators are Hermitian conjugates of each other and satisfy the commutation relation

$$[a, a^\dagger] = aa^\dagger - a^\dagger a = 1 . \tag{3.171}$$

This commutation relation shows that a is formally equivalent to $d/da^\dagger$. There is a vacuum state $|0\rangle$, containing no particles, which is annihilated by the action of a, so $a|0\rangle = 0$.

(i) If $\langle 0|0\rangle = 1$, what is $\langle n|n'\rangle$?

(ii) Show that $|n\rangle$ is an eigenstate of $N = a^\dagger a$.

(iii) What is $ae^{\lambda a^\dagger}|0\rangle$?

(iv) Show that $e^{\lambda a} f(a^\dagger) = f(a^\dagger + \lambda)e^{\lambda a}$.

(v) Show that

$$\int \frac{d3^2 z}{\pi} e^{-|z|^2} e^{za^\dagger}|0\rangle\langle 0|e^{z^* a} = 1 \tag{3.172}$$

(vi) Find the eigenvalues, right and left eigenstates of the operator

$$H = 2a^\dagger a^2 + a^{\dagger 2}a^2 .$$

(vii) Write down the Doi-Peliti hamiltonian in the shifted form for $A+A \to 0$ and $A + A \to A$. Show that they are related by a simple canonical transformation. What are the consequences for the asymptotic mean particle density in the two cases? Generalise your results to the processes $3A \to 0$, $3A \to A$ and $3A \to 2A$.

(viii) Show that we can define conjugate variables (θ, n), where $[\theta, n] = 1$, through the transformation

$$a = e^\theta n \qquad a^\dagger = e^{-\theta} . \tag{3.173}$$

Consider a model in which particles can hop between nearest neighbour sites $(j \to j+1)$ on a ring. The hopping rate $R_{j \to j+1}$ is a function $f(n_j)$ of the number of particles at site j before the hop. Write down the hamiltonian in terms of the above variables. Show that the stationary state (which is a right eigenstate of H with eigenvalue zero) has the product form $\prod_j p(n_j)$, and work out the function $p(n)$ in terms of $f(n)$. Is this the most general such hopping hamiltonian whose steady state has this simple product form?

(ix) For the reaction $2A \to 0$ considered in the lecture, with initial mean density n_0, draw some of the diagrams which contribute to the density-density correlation function $\overline{n(x,t)n(x',t')}$. Can you sum explicitly all the diagrams which contain no loops?

(x) For the reaction $3A \to 0$, draw the diagrams which renormalise the rate constant. What is the critical dimension? Evaluate the 1-loop correction to the rate constant renormalisation and hence sum all the vertex corrections to all orders. Evaluate the RG β-function and hence compute the asymptotic density in the critical dimension. [All these computations are a straightforward generalisation of the case $2A \to 0$ considered in lectures 3 & 4.]

(xi) Consider the irreversible reaction-diffusion process $A+B \to C$, where the products C of the reaction are inert. Suppose that a constant current $+J$ of A particles is imposed at $x = -\infty$, with an equal but opposite current $-J$ of B particles at $x = +\infty$, so that the system reaches a steady state. Write down the inhomogeneous rate equations for the steady state densities $a(x), b(x)$, (assumes that A and B have equal diffusion constants) and show that the reaction rate $R(x) = \lambda a(x)b(x)$ has a scaling solution as a function of x, D, λ and J. How does the width of the reaction zone scale with J in this approximation? What is the upper critical dimension d_c for the fluctuations in this problem? Without doing any explicit calculation, use RG ideas to argue how this scaling should be modified for $d < d_c$.

(xii) Prove that within Doi formalism, the moments of bosonic annihilation operator are equal to the *factorial* moments of the operator of the number of particles. Re-state your result in terms of solutions to the equivalent Langevin equation.

3.8.2 Solutions

(i) Solution:

$$\langle n|n' \rangle = \langle 0|a^n a^{\dagger n'}|0\rangle$$
$$= \langle 0|\frac{d^n}{da^{\dagger n}}a^{\dagger n'}|0\rangle \tag{3.174}$$

This is zero if $n' \neq n$ since there will either be surplus derivatives which annihilate the vacuum ket, or surplus creation operators which annihilate the vacuum bra. If $n = n'$ then we have

$$\langle n|n' \rangle = \delta_{n,n'}n!\langle 0|0\rangle = \delta_{n,n'}n! \tag{3.175}$$

This result can also be derived by commuting each a operator through the creation operators.

(ii) Solution:

$$a^\dagger a |n\rangle = a^\dagger \frac{d}{da^\dagger}(a^{\dagger\,n}|0\rangle)$$
$$= n(a^{\dagger\,n}|0\rangle)$$
$$= n|n\rangle . \tag{3.176}$$

(iii) Solution:

$$ae^{\lambda a^\dagger}|0\rangle = \frac{d}{da^\dagger}(1 + \lambda a^\dagger + \ldots + \frac{(\lambda a^\dagger)^m}{m!} + \ldots)|0\rangle$$
$$= (\lambda + \ldots + \frac{\lambda^m a^{\dagger\,m-1}}{(m-1)!} + \ldots)|0\rangle$$
$$= \lambda(1 + \lambda a^\dagger + \ldots + \frac{(\lambda a^\dagger)^{m-1}}{(m-1)!} + \ldots)|0\rangle$$
$$= \lambda e^{\lambda a^\dagger}|0\rangle . \tag{3.177}$$

(iv) Solution: First note that

$$e^{\lambda a} a^\dagger = (1 + \lambda a + \ldots)a^\dagger$$
$$= (1 + \lambda\frac{d}{da^\dagger} + \ldots)a^\dagger$$
$$= (a^\dagger + \lambda)(1 + \lambda\frac{d}{da^\dagger} + \ldots)$$
$$= (a^\dagger + \lambda)e^{\lambda a} . \tag{3.178}$$

So, by repeating this process we see that

$$e^{\lambda a} a^{\dagger\,n} = (a^\dagger + \lambda)^n e^{\lambda a} . \tag{3.179}$$

Thus the identity $e^{\lambda a} f(a^\dagger) = f(a^\dagger + \lambda)e^{\lambda a}$ holds for any function $f(a^\dagger)$ defined by a formal power series.

(v) Change to polar variables in the complex plane $z = \rho e^{i\theta}$

$$\int \frac{d^2 z}{\pi} e^{-|z|^2} e^{za^\dagger}|0\rangle\langle 0|e^{z^* a} = \int_0^{2\pi}\int_0^\infty \frac{\rho d\rho d\theta}{\pi} e^{-\rho^2} \sum_{m,n} \frac{\rho^{m+n} e^{i\theta(m-n)}}{m!n!} a^{\dagger\,m}|0\rangle\langle 0|a^n$$

The θ integral may be performed, giving 2π if $n = m$ and 0 otherwise, so

$$\int \frac{\mathrm{d}^2 z}{\pi} e^{-|z|^2} e^{za^\dagger} |0\rangle \langle 0| e^{z^* a} = 2\delta_{m,n} \int_0^\infty \rho \mathrm{d}\rho e^{-\rho^2} \sum_{m,n} \frac{\rho^{m+n}}{m!n!} a^{\dagger m} |0\rangle \langle 0| a^n$$

$$= \sum_n \frac{a^{\dagger n}}{\sqrt{n!}} |0\rangle \langle 0| \frac{a^n}{\sqrt{n!}}$$

$$= 1, \tag{3.180}$$

since

$$\int_0^\infty 2\rho \mathrm{d}\rho e^{-\rho^2} \frac{\rho^{2n}}{n!} = 1, \tag{3.181}$$

and from question (1), we see that $1/\sqrt{n!}$ is the correct normalization for the states.

(vi) Solution: The eigenvalue equation $H|\psi\rangle = E|\psi\rangle$ may be rewritten as a differential equation

$$(2x + x^2) \frac{\mathrm{d}^2}{\mathrm{d}x^2} f(x) = E f(x). \tag{3.182}$$

This may be solved with a Frobenius solution $f(x) = \sum_n a_n x^{n+l}$. Substituting this leads to

$$\sum_n 2a_n (n+l)(n+l-1)x^{n+l-1} + a_n(n+l)(n+l-1)x^{n+l} = \sum_n a_n E x^{n+l} \tag{3.183}$$

Equating powers of x, we make the choice $l = 1$ and obtain the requirement that

$$a_{n+1} = \frac{E - n(n+1)}{2(n+1)(n+2)} a_n. \tag{3.184}$$

The series must terminate so that the a_n do not become negative, which is unacceptable as probabilities must be positive. This termination happens, if $E = p(p-1)$, at the power x^p. The left eigenstates may be found from the Hermitian conjugate of the Hamiltonian with the same eigenvalues:

$$x^2 \left(2\frac{\mathrm{d}}{\mathrm{d}x} + \frac{\mathrm{d}^2}{\mathrm{d}x^2} \right) g(x) = p(p-1)g(x). \tag{3.185}$$

A Frobenius type solution, $g(x) = \sum_n a_n x^{n+l}$, may be attempted again.

$$\sum_n a_n 2(n+l)x^{n+l+1} + a_n(n+l)(n+l-1)x^{n+l} = \sum_n a_n p(p-1)x^{n+l}. \tag{3.186}$$

So for positive p, we require $l = p$ and find

$$a_{n+1} = \frac{-2(n+p)}{(n+2p)(1+n)} a_n \,. \tag{3.187}$$

The left eigenstates therefore start at level p and contain all higher powers of x.

(vii) Solution: The Hamiltonian for $A + A \to 0$ was derived from the master equation in the lectures. It takes the form

$$H = -\lambda(1 - a^{\dagger\,2})a^2 \,. \tag{3.188}$$

The Doi shifted form $(a^\dagger \to a^\dagger + 1)$ is

$$H_{\text{shifted}} = \lambda(2a^\dagger + a^{\dagger\,2})a^2 \,. \tag{3.189}$$

The master equation for $A + A \to A$ is

$$\frac{\mathrm{d}P(n)}{\mathrm{d}t} = \lambda(n+1)nP(n+1) - \lambda n(n-1)P(n) \,, \tag{3.190}$$

from which we obtain the Hamiltonian

$$H = -\lambda(b^\dagger - b^{\dagger\,2})b^2 \,. \tag{3.191}$$

The shifted Hamiltonian is seen to be

$$H = \lambda(b^\dagger + b^{\dagger\,2})b^2 \,. \tag{3.192}$$

The two shifted Hamiltonians are related by the canonical transformation

$$a^\dagger = \frac{1}{2}b^\dagger$$
$$a = 2b \,. \tag{3.193}$$

This has consequences for expectation values of operators, such as the number operator $a^\dagger a$.

$$\begin{aligned}
\overline{b^\dagger b}(t) &= \langle be^{-\lambda(b^\dagger + b^{\dagger\,2})b^2 t}e^b|\psi(0)\rangle \\
&= \langle be^{-\lambda(b^\dagger + b^{\dagger\,2})b^2 t}e^{n_0 b^\dagger}|0\rangle \\
&= \frac{1}{2}\langle ae^{-\lambda(2a^\dagger + a^{\dagger\,2})a^2 t}e^{2n_0 a^\dagger}|0\rangle \,.
\end{aligned} \tag{3.194}$$

Thus the average number of particles for the process $A + A \to A$ is twice that for the process $A + A \to 0$ with half as many particles in the initial state. The Hamiltonians for the three particle processes are

$$H(3A \to 0) = \lambda(a^{\dagger 3} - 1)a^3$$
$$H(3A \to A) = \lambda(b^{\dagger 3} - b^{\dagger})b^3$$
$$H(3A \to 2A) = \lambda(c^{\dagger 3} - c^{\dagger 2})c^3 \,. \tag{3.195}$$

The shifted forms are

$$H(3A \to 0) = \lambda(a^{\dagger 3} + 3a^{\dagger 2} + 3a^{\dagger})a^3$$
$$H(3A \to A) = \lambda(b^{\dagger 3} + 3b^{\dagger 2} + 2b^{\dagger})b^3$$
$$H(3A \to 2A) = \lambda(c^{\dagger 3} + 2c^{\dagger 2} + c^{\dagger})c^3 \,. \tag{3.196}$$

From these equations, it can be seen that a rescaling of the fields for one of the processes may be mapped onto a linear combination of the other two processes with rates λ_1 and λ_2.

(viii) Solution: If $[\theta, n] = 1$ then

$$[a, a^{\dagger}] = e^{\theta} n e^{-\theta} - 1 = (n+1) - n = 1 \,. \tag{3.197}$$

For later use, let us examine the action of $e^{\pm \theta}$ on a state $|\psi\rangle = \sum_n p(n)a^{\dagger n}|0\rangle$:

$$e^{-\theta}|\psi\rangle = \sum_n p(n)a^{\dagger n+1}|0\rangle = \sum_n p(n-1)a^{\dagger n}|0\rangle$$
$$e^{\theta}|\psi\rangle = \sum_n p(n)n^{-1}aa^{\dagger n}|0\rangle = \sum_n p(n)a^{\dagger n-1}|0\rangle = \sum_n p(n+1)a^{\dagger n}|0\rangle \,. \tag{3.198}$$

The hamiltonian for this process is a simple generalisation of the hopping hamiltonian considered in the lectures. It takes the form

$$H = \sum_j (a_{j+1}^{\dagger} - a_j^{\dagger})a_j f(n_j) \,, \tag{3.199}$$

which may be re-expressed in terms of the new variables as

$$H = \sum_j (e^{-\theta_{j+1}} - e^{-\theta_j})e^{\theta_j} n_j f(n_j) \,. \tag{3.200}$$

So, with reference to the results in equations 3.198, we see that the action of the hamiltonian on a state $|\psi\rangle$ is

$$H|\psi\rangle = \sum_j \left[\frac{p(n_{j+1} - 1)}{p(n_{j+1})} \frac{p(n_j + 1)}{p(n_j)}(n_j + 1)f(n_j + 1) - n_j f(n_j) \right]|\psi\rangle \,. \tag{3.201}$$

If we take the product state to be of the form

$$p(n_{j+1} - 1) = \lambda n_{j+1} f(n_{j+1}) p(n_{j+1}),$$ (3.202)

for all j, where λ is a constant, then the right hand side of equation 3.201 vanishes and the zero eigenstate is indeed of product form with

$$p(n) = \frac{p(n-1)}{\lambda n f(n)}.$$ (3.203)

There appear to be two independent parameters, $p(0)$ and λ, but in fact the constraint $\sum_n p(n) = 1$ means that there is only one. This parameter is the average density of particles on the ring.

(ix) Solution: Some of the diagrams contributing the density-density correlation function are shown in Fig 3.11.

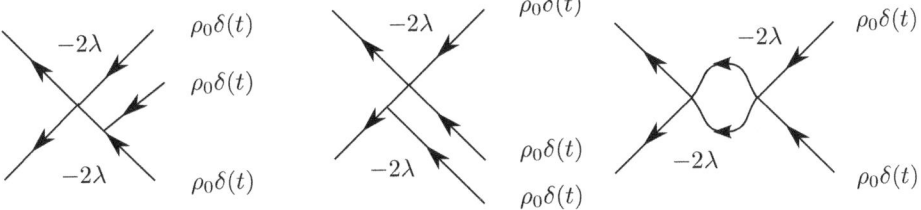

Fig. 3.11. Feynman diagrams contributing to the density-density correlation function.

These diagrams come from a perturbative expansion in the path integral representation. If $(x,t) \neq (x',t')$ then

$$n(x,t)n(x',t') = a_x^\dagger a_x a_{x'}^\dagger a_{x'} = a_x^\dagger a_{x'}^\dagger a_x a_{x'}.$$ (3.204)

The correlation function is therefore equal to $\langle \phi(x)\phi(x') \rangle$ in the path integral representation. The action for $2A \to 0$ is

$$S = \int \mathrm{d}^d x \int \mathrm{d}t \left[\tilde{\phi}(\partial_t - D_0 \nabla^2)\phi + 4\lambda \tilde{\phi} \frac{\phi^2}{2!} + 4\lambda \frac{\tilde{\phi}^2}{2!} \frac{\phi^2}{2!} - n_0 \tilde{\phi}(0) \right].$$ (3.205)

By rescaling time, D_0 may be set equal to 1, since the propagator is not renormalized for this action. Hence, Feynman diagrams are composed of a single propagator and a pair of vertices, as in fig 3.12. The factors proportional to λ in the vertex diagrams do not take symmetry factors into account, which are powers of 2 in diagrams associated with this action.

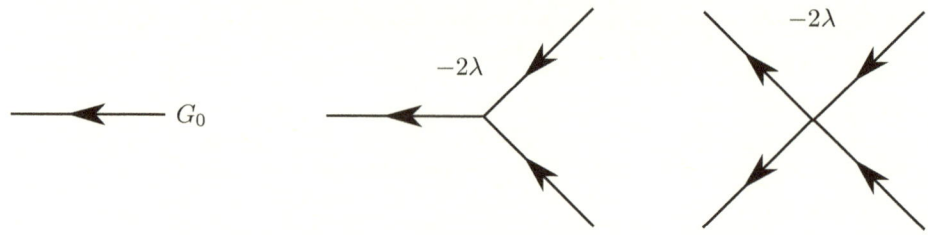

Fig. 3.12. The propagator and pair of vertices for $A + A \to 0$.

The density-density correlation function with no loops may be decomposed into two types of skeleton diagram. Firstly, the classical (no loops) density expectation diagram considered in the lectures, which is figure 3.13.

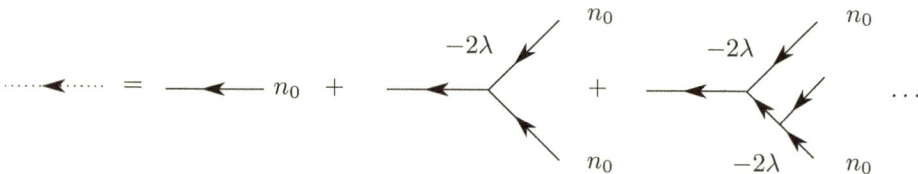

Fig. 3.13. The classical density expectation diagrams.

Note that the propagators from $t = 0$ carry zero momentum, since the initial state is spatially homogeneous. The propagator in (p, t) space is

$$G(p, t) = \int_{-\infty}^{\infty} \frac{1}{-i\omega + p^2} e^{-i\omega t} d\omega = e^{-p^2 t} \Theta(t) \,. \tag{3.206}$$

$\Theta(t)$ is the step-function.

$$\Theta(t > 0) = 1, \quad \Theta(t \le 0) = 0 \,. \tag{3.207}$$

The density expectation function is a geometric series in p, t space with all propagators equal to 1, so

$$n_{cl}(t) = \frac{n_0}{1 + 2n_0 \lambda t} \,. \tag{3.208}$$

The other type of diagram in the density-density correlation function is the response function $\langle \phi(-p, t_2) \tilde{\phi}(p, t_1) \rangle$, shown in Fig 3.14.

All classical density insertions are equivalent, so the time ordering of the vertices may be exchanged for a factor $1/n_v!$, where n_v is the

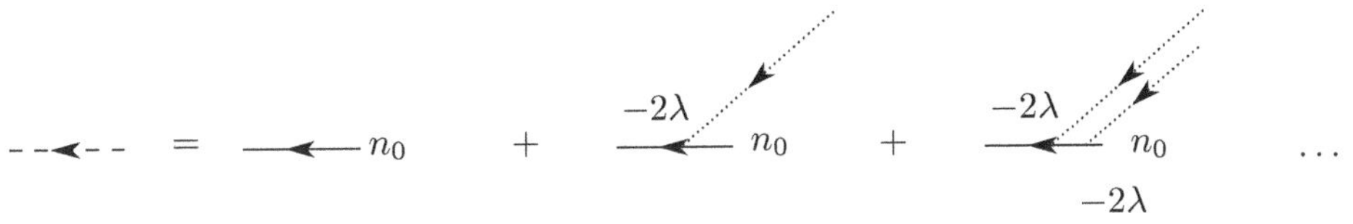

Fig. 3.14. The response function diagrams.

number of vertices in the diagram. The diagrams can then be seen
to correspond to the expansion of an exponential function

$$G(p, t_2, t_1) = e^{-p^2(t_2-t_1)}e^{-4\lambda \int_{t_1}^{t_2} n_{cl}(t')dt'}$$

$$= e^{-p^2(t_2-t_1)}\left(\frac{1 + 2n_0\lambda t_1}{1 + 2n_0\lambda t_2}\right)^2. \tag{3.209}$$

The density-density correlation function corresponds to the combination of response functions and classical density functions shown in
Fig 3.15.

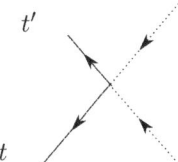

Fig. 3.15. The density-density correlation function.

This may be evaluated (assign $t' < t$) as

$$\overline{n(x,t)n(x',t')} = \frac{n_0^2}{(1 + 2n_0\lambda t)(1 + 2n_0\lambda t')}$$
$$- \int_{-\infty}^{\infty} \frac{e^{-i\mathbf{p}\cdot(\mathbf{x}-\mathbf{x'})}d^d\mathbf{p}}{(2\pi)^d} \frac{2\lambda n_0^2 e^{-p^2(t+t')}}{(1 + 2n_0\lambda t)^2(1 + 2n_0\lambda t')^2} \int_0^{t'} e^{2p^2 t''}(1 + 2n_0\lambda t'')^2 dt''. \tag{3.210}$$

Integrating over t'', the connected density-density correlation function
in (p, t) space is

$$\overline{n(p,t)n(-p,t')}_{\text{connected}} = -\frac{\lambda n_0^2 e^{-p^2(t+t')}}{(1 + 2n_0\lambda t)^2(1 + 2n_0\lambda t')^2 p^6}\left[-p^4 + 2n_0 p^2 \lambda\right.$$
$$\left. - 2n_0^2 \lambda^2 + e^{2p^2 t'}\left(2n_0^2\lambda^2 - 2n_0 p^2\lambda(1 + 2n_0\lambda t') + p^4(1 + 2n_0\lambda t')^2\right)\right]. \tag{3.211}$$

Note that if $(x, t) = (x', t')$, the commutation relation for ladder operators results in

$$\overline{n(x,t)n(x',t')} = \langle\phi(x,t)^2\rangle + \langle\phi(x,t)\rangle,\qquad(3.212)$$

so there is a correction equal to the classical density expectation function. The diffusion constant D_0 may be added into the above equations by replacing the propagators with $e^{-p^2 Dt}\Theta(t)$ and making the replacements $(\lambda, n_0) \to (D_0\lambda, D_0 n_0)$.

(x) Solution: The action for this model is similar to that of $2A \to 0$. In this case, it is

$$S = \int d^dx \int dt\left[\tilde\phi(\partial_t - D_0\nabla^2)\phi + 18\lambda\tilde\phi\frac{\phi^3}{3!} + 36\lambda\frac{\tilde\phi^2}{2!}\frac{\phi^3}{3!} + 36\lambda\frac{\tilde\phi^3}{3!}\frac{\phi^3}{3!} - n_0\tilde\phi(0)\right].$$
$$(3.213)$$

Rescaling time to remove D_0, we obtain the propagator and vertices shown in Fig 3.16.

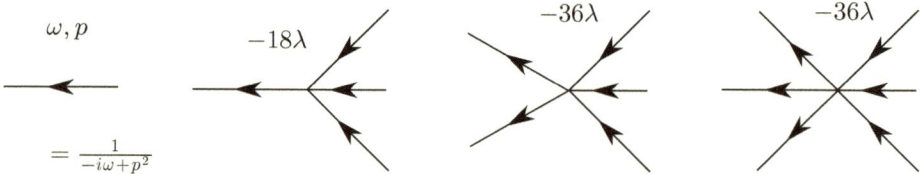

$$\omega, p$$

$$-18\lambda$$

$$-36\lambda$$

$$-36\lambda$$

$$= \frac{1}{-i\omega + p^2}$$

Fig. 3.16. The propagator and vertices for $3A \to 0$.

The dimension of the quantities of interest are

$$[\phi\tilde\phi] = p^d$$
$$[t] = p^{-2}$$
$$[\lambda] = p^{2-2d}.\qquad(3.214)$$

From this, we see that the critical dimension for $3A \to 0$ is $d_c = 1$, so we define $\epsilon = 1 - d$. The primitive diverges appear in the skeleton diagrams in Fig 3.17.

As for $2A \to 0$, the divergences in the three diagrams are produced by convolutions of the same bubble diagram. In this case the situation is similar, but now the divergences are contained in the two loop diagram of Fig 3.18.

This diagram contains the cube of the propagator in (x, t) space, which is

$$G(x, t) = \int_{-\infty}^\infty e^{-p^2t - i\mathbf{p}\cdot\mathbf{x}}\frac{d^d\mathbf{p}}{(2\pi)^d} = \frac{e^{-x^2/4t}}{(4\pi t)^{d/2}}.\qquad(3.215)$$

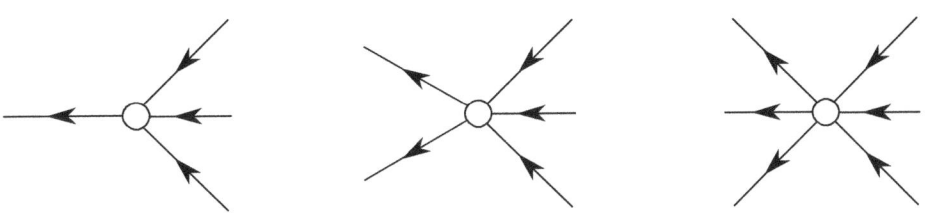

Fig. 3.17. Skeleton diagrams with primitive divergences for $3A \rightarrow 0$.

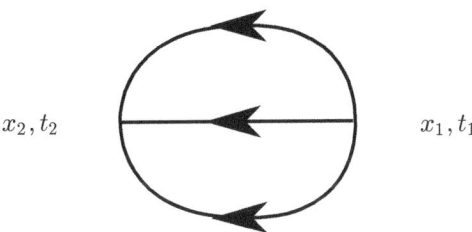

Fig. 3.18. Divergent two loop diagrams.

Although the bubble diagram is most straightforwardly expressed in (x, t) space, the series of diagrams are most easily evaluated in (p, s) space, since the Fourier or Laplace transforms of convolutions are products. Hence, we take the transforms of equation 3.215. First, the Fourier transform into p space

$$3!\lambda_0 \int_{-\infty}^{\infty} \frac{e^{-3x^2/4t}}{(4\pi t)^{3d/2}} e^{i\mathbf{p}\cdot\mathbf{x}} \mathrm{d}^d\mathbf{x} = \frac{3!\lambda_0}{3^{d/2}} \frac{1}{(4\pi(t_2 - t_1))^d} e^{-p^2(t_2 - t_1)/3} \,.$$

$$(3.216)$$

The factor 3! comes from combinatorics. Then, the Laplace transform into s space

$$\int_0^{\infty} \frac{3!\lambda_0}{3^{d/2}} \frac{1}{(4\pi(t_2 - t_1))^d} e^{-p^2(t_2 - t_1)/3} e^{-st} \mathrm{d}t = \frac{3!\lambda_0}{3^{d/2}(4\pi)^d} \Gamma(\epsilon)(s + p^2/3)^{-\epsilon} \,.$$

$$(3.217)$$

The sum of diagrams is therefore

$$\lambda_R(p, s) = \frac{\lambda_0}{1 + \frac{3!\lambda_0}{3^{d/2}(4\pi)^d} \Gamma(\epsilon)(s + p^2/3)^{-\epsilon}} \,. \qquad (3.218)$$

Using this, we may now renormalise the expected number of particles in the model. This follows the same line of reasoning as for $2A \rightarrow 0$, which was explained in the lectures.

$$s\frac{\mathrm{d}}{\mathrm{d}s}\Big|_{\lambda_0} \overline{n}(t, g_R, s) = 0 \,. \qquad (3.219)$$

Introduce the dimensionless coupling $g_R = \lambda_R s^{-\epsilon}$ so that

$$(s\frac{\partial}{\partial s} + \beta(g_R)\frac{\partial}{\partial g_R})\overline{n}(t, g_R, s) = 0, \tag{3.220}$$

with

$$\beta(g_R) = s\frac{\partial g_R}{\partial s}\Big|_{\lambda_0}. \tag{3.221}$$

In the critical dimension $d_c = 1$, the beta function is $\beta(g_R) = \sqrt{3}g_R^2/2\pi$. Using dimensional analysis, $\overline{n} = t^{-d/2}f(ts, g_R)$, so that the solution to the Callan-Symanzik equation is

$$\overline{n}(t, g_R, s) = (ts)^{-d/2}\overline{n}(t = s^{-1}, \tilde{g}(t), s), \tag{3.222}$$

where

$$t\frac{\mathrm{d}}{\mathrm{d}t}\tilde{g}(t) = -\beta[\tilde{g}(t)]. \tag{3.223}$$

In $d = 1$, this has solution

$$\tilde{g}(t) = \frac{2\pi}{\sqrt{3}\ln t}, \tag{3.224}$$

If we take $s \ll 1$, then we may use the classical asymptotic solution to the rate equation

$$\frac{\mathrm{d}\overline{n}}{\mathrm{d}t} = -3\lambda\overline{n}^3, \tag{3.225}$$

with solution

$$\overline{n} = \frac{1}{\sqrt{3!\lambda t}}. \tag{3.226}$$

Substituting this into equation 3.222, we find

$$\overline{n} = (st)^{-1/2}\frac{1}{\sqrt{3!\frac{2\pi}{\sqrt{3}\ln t}s^{-1}}} = \sqrt{\frac{\ln t}{4\sqrt{3}\pi t}}. \tag{3.227}$$

Hence, we see the logarithmic corrections to the expected particle number in the critical dimension, with a universal amplitude. In $2A \to 0$, we saw this behaviour in $d = 2$.

(xi) Solution: The action for this process must be dimensionless, since it appears as the argument of an exponential function. The dimensions of the various quantities are

$$[\tilde{\phi}_a] = [\tilde{\phi}_b] = [\tilde{\phi}_c] = 1$$
$$[\phi_a] = [\phi_b] = [\phi_c] = p^d$$
$$[D] = 1$$
$$[t] = p^{-2}$$
$$[\lambda] = p^{-d+2} \tag{3.228}$$

From equation 3.228, we see that the upper critical dimension for this process is $d_c = 2$. For $d > 2$, mean field theory is asymptotically correct. The rate equations are

$$\partial_t a = D\nabla^2 a - \lambda ab \tag{3.229}$$

$$\partial_t b = D\nabla^2 b - \lambda ab \,. \tag{3.230}$$

Subtracting the equations above, we see that

$$\nabla^2(a - b) = 0 \,. \tag{3.231}$$

The expected densities should only be a function of the distance parallel to the direction of the currents. Call this direction the x direction. Therefore,

$$\frac{d}{dx}(a - b) = \text{constant} = J \,, \tag{3.232}$$

where J is the current. This has solution $a - b = Jx$ which may be substituted back into equation 3.229:

$$\partial_t a = D\frac{d^2 a}{dx^2} - \lambda a^2 + \lambda Jxa \,. \tag{3.233}$$

Consider the transformed variables

$$t' = st$$
$$x' = lx$$
$$a' = ma \,. \tag{3.234}$$

Substituting these into equation 3.233

$$(sm^{-1})\partial_{t'}a' = D(l^2 m^{-1})\frac{d^2 a'}{dx^2} - \lambda(m^{-2})a'^2 + \lambda J(l^{-1}m^{-1})x'a' \,. \tag{3.235}$$

For the correct choice of $s, l \,\&\, m$, the quantities $\lambda, D \,\&\, J$ may be scaled out. We require

$$l^2 m^{-1} = KD^{-1}$$
$$sm^{-1} = K$$
$$m^{-2} = K/\lambda$$
$$l^{-1}m^{-1} = K/\lambda J \,. \tag{3.236}$$

Solving these equations, we find

$$K = \lambda^{1/3} J^{4/3} D^{2/3}$$
$$l = \lambda^{1/3} J^{1/3} D^{-1/3}$$
$$m = \lambda^{1/3} J^{-2/3} D^{-1/3}$$
$$s = \lambda^{2/3} J^{2/3} D^{1/3} \,. \tag{3.237}$$

With these choices for $K, l, m \, \& \, s$, the transformed variables satisfy the equation

$$\partial_{t'} a' = \frac{\mathrm{d}^2 a'}{\mathrm{d}x^2} - a'^2 + x'a' \,, \tag{3.238}$$

so the steady-state density, a', is a function only of x'. From this, we may obtain a scaling form for $a(x)$, since it should be related to $a'(x')$ by

$$a(x) = m^{-1} a'(lx)$$
$$= J^{2/3} \left(\frac{D}{\lambda} \right)^{1/3} \left(J^{1/3} \left(\frac{\lambda}{D} \right)^{1/3} x \right) . \tag{3.239}$$

So, for fixed λ and D, the density has the scaling form:

$$a(x) \sim J^{2/3} \psi(J^{1/3} x) \,, \tag{3.240}$$

where ψ is some unknown function. Similarly,

$$b \sim J^{2/3} \phi(x J^{1/3}) \,, \tag{3.241}$$

so the rate term, $R = \lambda ab$ scales as

$$R = \lambda ab \sim J^{4/3} \chi(x J^{1/3}) \,. \tag{3.242}$$

The width of the reaction zone must scale as all lengths in this problem for fixed D, so $w \sim J^{-1/3}$.

Now let us consider the case $d < d_c$. We expect the behaviour to be non-mean field, but not to depend on λ, since this will flow to a fixed point of the beta function. $\langle a \rangle$ should not depend on the artificially introduced renormalization scale s, so

$$s\frac{\mathrm{d}}{\mathrm{d}s}\Big|_{\lambda_0}\langle a \rangle(x, J, D, g_R(s), s) = 0. \tag{3.243}$$

Expand this derivative into a term which acts explicitly on the s dependence of $\langle a \rangle$ and a term acting on the $g_R(s)$ piece.

$$\Big(s\frac{\partial}{\partial s} + \beta g_R\frac{\partial}{\partial g_R}\Big)\langle a \rangle(x, J, D, g_R(s), s) = 0. \tag{3.244}$$

We make use of dimensional analysis to fix a form for $\langle a \rangle$. The dimensions of the various quantities are

$$[a] = p^d$$
$$[x] = p^{-1}$$
$$[D] = 1$$
$$[t] = [x^2] = p^{-2}$$
$$[s] = [t^{-1}] = p^2$$
$$[J] = \frac{[a]}{[x]} = p^{d+1} \tag{3.245}$$

So,

$$\langle a \rangle = x^{-d} f(x s^{1/2}, J s^{-(d+1)/2}, D, g_R, s), \tag{3.246}$$

and the Callan Symanzik equation may be written as

$$\Big(x\frac{\partial}{\partial x} + d - (d+1)J\frac{\partial}{\partial J} + 2\beta(g_R)\frac{\partial}{\partial g_R}\Big)\langle a \rangle(x, J, D, g_R(s), s) = 0. \tag{3.247}$$

We could solve this via the method of characteristics, but to get the aymptotic behaviour for $d < 2$ we can simply set $\beta(g_R) = 0$. The scaling form for the solution is then

$$\langle a \rangle = J^{d/(d+1)}\,\psi\big(x J^{1/(d+1)}\big). \tag{3.248}$$

Note that the scaling now depends on d. We could have got this by simply assuming that the asymptotic behaviour is independent of λ, and using dimensional analysis, but the RG approach systematises this.

To get the scaling for R, we can't use the mean field relation $R = \lambda\langle a \rangle\langle b \rangle$ any more. Instead note that it has the same dimensions as $\dot{a}$, that is p^{d+2}. Repeating the same arguments as above, we then find

$$R = J^{(d+2)/(d+1)}\,\chi\big(x J^{1/(d+1)}\big). \tag{3.249}$$

This tells us that in $d = 1$ the height and width of the reaction zone scale like $J^{3/2}$ and $J^{-1/2}$ respectively, quite different from the mean field results.

(xii) Solution: To simplify the discussion, we consider 0-dimensional case first. Let $Z_t(J)$ be the generating function of moments for the probability distribution $P_t(N)$. By definition,

$$Z_t(J) \equiv E(e^{JN_t}) = \sum_0^\infty e^{JN} P_t(N).$$

In terms of bosonic ladder operators,

$$Z_t(J) = \langle 0|e^a e^{Ja^\dagger a}|P_t\rangle. \tag{3.250}$$

The form of the correlation function on the right hand side of eq. (3.250) is not very suitable for practical computations due to the presence of the shift operator e^a inside the brackets. This problem can be overcome by commuting e^a to the right and using $e^a|0\rangle = |0\rangle$. It follows from the bosonic commutation relation $[a, a^\dagger] = 1$ that $e^a O(a^\dagger, a) = O(a^\dagger + 1, a)e^a$ for any operator $O(a^\dagger, a)$. Using this fact one finds that

$$Z_t(J) = \langle 0|e^{J(a^\dagger + 1)a}e^{-t\tilde{H}}|\tilde{P}_0\rangle, \tag{3.251}$$

where $\tilde{H}(a^\dagger, a) = H(a^\dagger + 1, a)$ is the shifted Hamiltonian, $|\tilde{P}_0\rangle = e^a|P_0\rangle$. The expression $\langle 0|e^{J(a^\dagger + a)a}$ can be simplified further. Note that $[a^\dagger a, a] = -a$. In other words, the operators a and $a^\dagger a$ form a basis of a Lie algebra isomorphic to a subalgebra of $sl(2)$ consisting of upper triangular matrices. Consequently, the Campbell-Hausdorff formula implies that

$$e^{J(a^\dagger + a)a} = e^{Ja^\dagger a}e^{f(J)a}, \tag{3.252}$$

where $f(J)$ is a function to be determined. Differentiating both sides of eq. (3.252) with respect to J, commuting all operators multiplying exponents in the derivatives to the right, and comparing both sides of the resulting equality, we find a differential equation for f:

$$f'(J) = f(J) + 1, \tag{3.253}$$

which should be solved with the boundary condition $f(0) = 0$. The answer is $f(J) = e^J - 1$. Taking into account that $\langle 0|e^{Ja^\dagger a} = \langle 0|$, we find

$$Z_t(J) = \langle 0|e^{(e^J - 1)a}e^{-t\tilde{H}}|\tilde{P}_0\rangle, \tag{3.254}$$

Let $\theta = e^J - 1$. Differentiating the above equation with respect to θ k times and setting $\theta = 0$ we find that

$$E\big(N_t(N_t - 1)(N_t - 2)...(N_t - k + 1)\big)_{P_t}$$
$$= \langle a(t)^k \rangle_{LE}, k = 1, 2, 3, \ldots \qquad (3.255)$$

where the averaging in the r.h.s. is with respect to stochastic noise in Langevin equation satisfied by $a(t)$ (Recall that the field $a(t)$ corresponds to the bosonic operator a in Doi's formalism.)

Therefore, the factorial moments of the p. d. f. of the number of particles $P_t(N)$ are equal to monomial moments of the solution to Langevin equation $a(t)$. If the expected number of particles is small, the main contribution to the left hand side of (3.255) comes from a k-particle configuration. Therefore,

$$k!Prob(N_t = k) \approx \langle a(t)^k \rangle_{LE}, k = 1, 2, 3, \ldots, \qquad (3.256)$$

which relates the p. d. f. of the particle number N_t to moments of $a(t)$.

The generalization of this result to $d > 0$ is straightforward: in the limit of low particle density,

$$Prob\big(\vec{x}_1, \ldots, \vec{x}_n \in \Delta V(\vec{x})\big) \approx \frac{1}{n!} E\big(\Delta\phi^n(\vec{x}, t)\big)_{LE} \qquad (3.257)$$

where

$$\Delta\phi_t(\vec{x}) = \int_{\Delta V(\vec{x})} d^d x' \phi_t(\vec{x}'), \qquad (3.258)$$

and $\phi_t(\vec{x})$ is a solution of a d-dimensional Langevin equation with imaginary multiplicative noise (stochastic rate equation).

Bibliography

[Tauber(2007)] U.C. Tauber, *Critical dynamics: A field theory approach to equilibrium and non-equilibrium scaling behavior*, in preparation, Cambridge University Press; for completed chapters, see
$http://www.phys.vt.edu/ \sim tauber/utaeuber.html$.

[Cardy(1997)] J.L. Cardy, *Renormalisation group approach to reaction-diffusion problems*, in: Proceedings of Mathematical Beauty of Physics, Ed. J.-B. Zuber, Adv. Ser. in Math. Phys. **24**, 113 (1997); cond-mat/9607163.

[Tauber et al.(2005)] U.C. Tauber, M.J. Howard, and B.P. Vollmayr-Lee, *Applications of field-theoretic renormalization group methods to reaction-diffusion problems*, J. Phys. A 38, R79 (2005); cond-mat/0501678.

[Frisch(1995)] U. Frisch, *Turbulence : the Legacy of A.N. Kolmogorov* (Cambridge University Press, Cambridge, 1995).

[Falkovich et al.(2001)] G. Falkovich, K. Gawedzki, and M. Vergassola, Rev. Modern Phys. **73**, 913 (2001).

[Zakharov et al.(1992)] V. Zakharov, V. Lvov, and G. Falkovich, *Kolmogorov Spectra of Turbulence* (Springer-Verlag, Berlin, 1992).

[Connaughton et al.(2005)] C. Connaughton, R. Rajesh, and O. Zaboronski, Phys. Rev. Lett. **94**, 194503 (2005).

[R. Rajesh et al.(2006)] C. Connaughton, R. Rajesh and O. Zaboronski, *Cluster-Cluster Aggregation as an Analogue of a Turbulent Cascade : Kolmogorov Phenomenology, Scaling Laws and the Breakdown of self-similarity* , cond-mat/0510389, Physica D, **Vol 222**, 97 (2006) ;

[Scheidegger(1967)] A. E. Scheidegger, Bull. IASH. **12**, 15 (1967).

[Dodds and Rothman(1999)] P. S. Dodds and D. H. Rothman, Phys. Rev. E **59**, 4865 (1999).

[Takayasu(1989)] H. Takayasu, Phys. Rev. Lett. **63**, 2563 (1989).

[Huber(1991)] G. Huber, Physica A **170**, 463 (1991).

[Rajesh and Majumdar(2000)] R. Rajesh and S. Majumdar, Phys. Rev. E **62**, 3186 (2000).

[Connaughton et al.(2004)] C. Connaughton, R. Rajesh, and O. Zaboronski, Phys. Rev. E **69**, 061114 (2004).

[Connaughton et al.(2007)] C. Connaughton, R. Rajesh and O. Zaboronski, *Constant Flux Relation for Driven Dissipative Systems*, cond-mat/0607656, PRL, **Vol 98**, 080601 (2007).

[Krapivsky et al.(1999)] P. L. Krapivsky, J. F. F. Mendes, and S. Redner, Phys.

Rev. B **59**, 15950 (1999).

[Coppersmith et al.(1996)] S. N. Coppersmith, C. h Liu, S. Majumdar, O. Narayan, and T. A. Witten, Phys. Rev. E **53**, 4673 (1996).

[Majumdar et al.(2000)] S. N. Majumdar, S. Krishnamurthy, and M. Barma, Phys. Rev. E **61**, 6337 (2000).

[Rajesh(2004)] R. Rajesh, Phys. Rev. E **69**, 036128 (2004).

[Dhar and Ramaswamy(1989)] D. Dhar and R. Ramaswamy, Phys. Rev. Lett. **63**, 1659 (1989).

[Doi(1976a)] M. Doi, J. Phys. A **9**, 1465 (1976a).

[Doi(1976b)] M. Doi, J. Phys. A **9**, 1479 (1976b).

[Cardy()] J. Cardy, *Field theory and nonequilibrium statistical mechanics*, notes are available at the website http://www-thphys.physics.ox.ac.uk/users/JohnCardy/.

[Lee(1994)] B. Lee, J. Phys. A **27**, 2633 (1994).

[Zaboronski(2001)] O. Zaboronski, Phys. Lett. A **281**, 119 (2001).

[Martin et al.(1973)] P. Martin, E. Siggia, and H. Rose, Phys. Rev. A **8**, 423 (1973).

[Peliti(1986)] L. Peliti, J. Phys. A **19**, L365 (1986).

[Droz and Sasvári(1993)] M. Droz and L. Sasvári, Phys. Rev. E **48**, R2343 (1993).

[Binney et al.(1992)] J. Binney, N. Dowrick, A. Fisher, and M. Newman, *The Theory of Critical Phenomena* (Clarendon Press, Oxford, 1992), chap. 11 contains an excellent pedagogical introduction to the renormalisation group in the context of the Landau-Ginzberg model, much of which is equally applicable to the problem under study here.

[Krapivsky et al.(1994)] P. Krapivsky, E. Ben-Naim, and S. Redner, Phys. Rev. E **50**, 2474 (1994).

[Kontorovich(2001)] V. Kontorovich, Physica D **152-153**, 676 (2001).

[Biven et al.(2001)] L. Biven, S. Nazarenko, and A. Newell, Phys. Lett. A **280**, 28 (2001).

For EU product safety concerns, contact us at Calle de José Abascal, 56–1°,
28003 Madrid, Spain or eugpsr@cambridge.org.

www.ingramcontent.com/pod-product-compliance
Ingram Content Group UK Ltd.
Pitfield, Milton Keynes, MK11 3LW, UK
UKHW010854090126
466816UK00011B/241